INTERSCIENCE TRACTS
IN PURE AND APPLIED MATHEMATICS

Editors: L. BERS · R. COURANT · J. J. STOKER

Additional volumes in preparation

INTERSCIENCE TRACTS
IN PURE AND APPLIED MATHEMATICS

Editors: L. BERS · R. COURANT · J. J. STOKER

Number 21

MATHEMATICAL STRUCTURES OF LANGUAGE
By Zellig Harris

INTERSCIENCE PUBLISHERS
a division of John Wiley & Sons   New York London Sydney Toronto

MATHEMATICAL
STRUCTURES
OF LANGUAGE

ZELLIG HARRIS

University of Pennsylvania
Philadelphia, Pennsylvania

INTERSCIENCE PUBLISHERS

A DIVISION OF JOHN WILEY & SONS, NEW YORK · LONDON · SYDNEY · TORONTO

Library of Congress Catalog Card Number: 68-16053

SBN 470   353163

Printed in the United States of America

Preface

This book attempts to show how one can arrive at an abstract system which characterizes precisely natural language. This is done by taking the data of language and finding within the data such relations as can be organized into a suitable model. The problem here was not to find a broad mathematical system in which the structure of language could be included, but to find what relations, or rather relations among relations, were necessary and sufficient for language structure. Then anything which is an interpretation of the model can do the work of natural language. Modifications of the abstract system are then considered, which serve as models for language-like and information-bearing systems.

Since this book is an expansion of a lecture given at the Courant Institute of Mathematical Sciences, it reports only the writer's work, and does not attempt a general survey of what has been called mathematical linguistics. Such a survey could have no unified form at present, since the field is not yet an established one with interrelated problems and methods.

Z. S. HARRIS

June 1968

Contents

1

Introduction

Mathematical linguistics, as developed here, characterizes natural language as a system of sets of arbitrary objects, the sets being closed with respect to particular operations, with certain mappings of these sets into themselves or into or onto related sets. The operations and mappings have interpretations which yield the meanings of the utterances of the language.

The only body of data required for the whole analysis of language is the indication that certain sound sequences, out of some large sample, are utterances of the language (with normal acceptance, or less) while others are not, and that certain ones are repetitions of each other. Structural linguistics shows how these utterances can be characterized as a set of constructions on certain discrete elements. Mathematical linguistics shows that the characterization can be made in terms of other sets, defined by certain relations among these linguistic elements, and that the entities in the new sets are arbitrary and are defined only by the relations among the new sets.

The interest here is not in investigating a mathematically definable system which has some relation to language, as being a generalization or a subset of it, but in formulating as a mathematical system all the properties and relations necessary and sufficient for the whole of natural language.

The main result so far has been in defining an abstract system that fits language, and in constructing a few further systems as extensions of this, rather than in proving theorems about the abstract systems. It may be expected, however, that theorems which are proved about these systems would be interpretable as true properties of language, since the systems are built to describe neither more nor less than natural language.

On the basis of these formulations it is possible to characterize the set of utterances—discourses or sentences—of a language and many interesting subsets of that set. Sentences are shown to have, aside from certain degeneracies, a unique and computable factorization into prime sentences; equivalently, there is a recursive method for obtaining all sentences from a finite elementary subset (of assertions) by means of a relatively small set of operators. The properties of each sentence are completely accounted for

1

by the properties of its factorization. There are also effective methods for constructing various languagelike systems with predictable properties (including languages without ambiguity, without synonyms, etc.) and for specifying the relevant differences between natural language and logic or mathematics.

The system of operators turns out to have a simple structure which fits the work that language does. It is therefore not surprising that this system is much the same for all languages that have been investigated from this point of view. The relevance of the system is further supported by the fact that even irregularities of the grammar are found to be special cases of regular operations. Furthermore, the primitive elements—elementary sentences K and operators ϕ—have meanings, i.e., a linguistic interpretation, such that the meaning of a sentence, as a sequence of K and ϕ, is the sequence of meanings of its component K and ϕ. No major independent semantic theory is thus needed.

Since neither the physical properties nor yet the meanings of the sounds and words are used in determining the structure, the linguistic meanings which the structure carries can only be due to the relations in which the elements of the structure take part. Thus any set of objects which have the operations and mappings described here can be interpreted as a natural language, bearing the kind of meanings or information that language carries. Modifications of the abstract system would then, under the linguistic interpretation, characterize particular modifications, generalizations, and specializations of natural language (such as a language of science). But the possibility also arises of seeking other interpretations than language for the abstract system.

The kinds of mathematical structures which are found adequate for language may thus have a general interest, and the purpose of the present study has been to determine what these structures are. The system which is finally obtained (Chapter 7) consists of one family of primitive arguments and five finite families of operators, each defined as operating on the primitive arguments or on operators. Each family has a structure of its own. Each entity (primitive argument or operator) brings one word (or affix) into the sentence which is being produced; or in a few cases brings a sequence or a change in the existing words. The system itself is thus rather special—it could hardly be otherwise—and may or may not be interesting to investigate in itself. In any case, it may be interesting to compare it with its nearest mathematical neighbors, something which is possible once it has been established.

The abstract system and its linguistic interpretation show that the composition and meaning of sentences are invariant under certain of the operators. Hence the ϕ can be distinguished into incremental ϕ which construct the meaning of the sentence and deformational ϕ which leave the meaning unchanged. We can then construct a simpler abstract system, consisting of a finite set of primitive arguments and two finite sets of operators, which expresses all the information carried by natural language, and in which the symbol sequences vary in a regular way precisely as the information varies.

It has not been possible to construct structures adequate for language by means of a few simple operations on the directly observable objects of natural language, such as the sounds and the words. Little can be accomplished with the directly observable elements, as compared with what can be done if we take as our elements certain suitable relations among the observables. Thus, the set $\{S\}$ of sentences, taken only as sequences of words, is apparently not characterizable in any unified theory. When the primitives are certain short strings of words, determined by specified relations among word-occurrences in utterances, $\{S\}$ is obtainable by a recurrent hereditary stochastic process (3.2, 6). When the primitives are the ϕ, K of Chapter 4, each of which is a short sequence or change of words which is determined by definite criteria, $\{S\}$ is describable by a Markov chain. The great bulk of the work here has been in finding how to determine such subsets (i.e., relations) of utterances as would constitute adequate objects for relevant operations and mappings. In order to show something of the kind of work that is involved in worrying the data of language into this form, and in order to give some evidence that the structures proposed here are indeed referable to the actual data of language (and in principle to all of it), it was necessary to include a certain amount of linguistic detail (especially in 4.2.2), which is not otherwise of general interest, especially in this condensed form. Since the mathematically defined structures are obtained from observable relations within the data, this book becomes in a way a sketch of linguistic theory. But since many of the considerations and results which make for a desirable theory were reached via the mathematical formulation of the structures, this is as much a book about the use of mathematical constructions as about the organizing of linguistic data. However, the characterization of language reached here has not been presented as a formal theory. There are various problems still open as to what can best be taken as the axioms of such a theory, and what should be taken as proved from them.[1]

---

[1] Since it may be supposed that various chapters of this book may be used separately, some attempt has been made to keep the chapters independent of each other, by occasional summarizing of the relevant material from preceding chapters.

The work summarized here[2] is presented with greater linguistic detail in a number of earlier papers and monographs, chiefly:

For structural and string theories:
Methods in structural linguistics, University of Chicago Press, 1951.
From phoneme to morpheme, Language 31 (1955) 190–222.
String analysis of sentence structure, Mouton, The Hague, 1962.
A cycling cancellation-automaton for sentence well-formedness, Bulletin International Computation Centre 5 (1966) 69–94.

For transformations and discourse analysis: the first presentation of linguistic transformations is in pp. 18–25 of Discourse Analysis, Language 28 (1952) 1–30, followed by
Co-occurence and transformation in linguistic structure, Language 33 (1957) 283–340.
Transformational theory, Language 41 (1965) 363–401.
Discourse analysis reprints, Mouton, The Hague, 1963.

The general method of phonemic analysis (3.1) is standard in modern linguistics. The major original works are:
F. de Saussure, Cours de linguistique generale, Paris, 1910.
Edward Sapir, Sound patterns in language, Language 1 (1925) 37–51.
Leonard Bloomfield, Language, Holt, 1933.

In the subject of transformational theory, and in another direction of mathematical linguistics, special reference should be made to the work of Noam Chomsky, who has combined transformational analysis with a generative theory of sentence structure, and has also produced a mathematical specification of context-free languagelike systems within a spectrum of languagelike systems. See chiefly:

N. Chomsky, Syntactic structures, Mouton, The Hague, 1957.
——, Three models for the description of language, IRE Transactions on information theory, IT-2 (1956) 133–24.
——, A transformational approach to syntax, in A. A. Hill, Proceedings of the third Texas conference on problems of linguistic analysis in English, 1958. (Univ. of Texas, 1962).
——, On certain formal properties of grammars, Information and control 2 (1959) 137–67.
——, On the notion "Rule of grammar," Proceedings Symposium in applied mathematics 12. 6–24, American Mathematical Society, 1961.

---

[2] The present text has had the advantage of valuable comments from Henry Hiż and Aravind K. Joshi, for which I am glad to thank them.

————, Aspects of the theory of syntax, MIT Press, 1965.

Mathematical aspects of essential linguistic properties have been studied by M. P. Schützenberger, as in:

N. Chomsky and M. P. Schützenberger, The algebraic theory of context-free languages, in Braffort and Hischberg, Computer programming and formal systems, North-Holland, Amsterdam, 1963.

The methods of mathematical logic have been brought to bear upon these problems primarily in the work of Henry Hiż, as in:

H. Hiż, The intuition of grammatical categories, Methodos, 1961.

————, Congrammaticality, batteries of transformations, and grammatical categories, Proceedings Symposium in applied mathematics 12. 43–50, American Mathematical Society, 1961.

————, The role of paraphrase in grammar, Monograph series on Language and Linguistics, vol. 17, Georgetown University, Washington, D.C., 1964.

————, Disambiguation. To appear in the Proceedings of the Colloque International de Sémiologie Kazimierz, 1966. Also as Transformations and Discourse Analysis Papers, 1966.

2

Properties of language relevant to a mathematical formulation

2.0.

Before attempting any mathematical formulation of natural language, we consider certain apparently universal and essential properties of language, which are observable without any mathematical analysis, and which are such as to make possible a mathematical treatment. In this chapter we refer to the characteristic phenomena of linguistics and to the elements which have been successfully set up in it (phonemes, morphemes, grammatical intonations, etc.), and not to any entities defined later by means of mathematical types of analysis.

2.1. Elements are discrete, preset, arbitrary

In the analysis of language, it turns out that the only elements which enter into grammatical structure are discrete ones. These discrete elements of grammar are in some cases obtained from continuous physical events: The discrete elements are defined as cuts in sets of continuous events.[1] For example, certain differences among the continuously varying sound events of various utterances become the basis for defining the phonemic elements (3.1); in a different way, transformations are defined with the aid of an ordering of the degree of acceptability of sentences (4.1). When continuous phenomena appear in grammatical relations, it is only after they have been organized into discrete elements which have been defined on the basis of differences among the continuous phenomena. Those linguistic phenomena which are not represented by discrete elements may indeed carry information (e.g., intonations of hestitation or of exaggerated matter-of-factness), but they do not enter into grammatical relations to other elements.

The utilization of solely discrete elements is uniquely appropriate for reasonably error-free transmission of utterances. There has been much discussion as to whether language is an instrument of expression or of communication. Instead, its structure is peculiarly suited for an instrument of transmission. Expression and face-to-face communication are served quite well by such continuous phenomena as intonations, phonetic characteristics of individual speakers, gestures, etc.; and the hearer may

[1] The discreteness discussed here is over and above the discreteness which is due to physical limitations of the speaker's movements and of processing in the brain.

6

garner a great deal of information from such continuous data, over and above what is grammatically structured in the speaker's utterance. However, when the spoken material is transmitted, i.e., spoken by the hearer to a new hearer, and so down the line, there is a possibility of compounding errors in one direction when continuous elements are transmitted. In contrast, when the utterance is characterized by discrete elements only, it can tolerate a known and considerable amount of noise in transmission from one hearer to the next without producing error. Grammatical utterances are, in fact, rather awkward carriers of expression and face-to-face communication: feelings are expressed in language not directly but by making a statement that one has those feelings; and many nuances of perception cannot be distinguished in language. What is special to a grammatical utterance (i.e., to a linguistic event) is not that it has meaning,[2] expresses feelings, communicates, or calls for a relevant response—these are all common to many human activities—but that it is socially transmissible.[3]

Transmissibility without error compounding requires not only that the elements of the material to be transmitted be discrete, but also that they be preset in sender and receiver. When the speaker and the hearer are referring to a set of elements known to both, the hearer need receive only enough of a signal to distinguish a particular element in contrast to all other elements. When the hearer then transmits (repeats) the utterance, he pronounces his own rendition of the preset (i.e., known) elements which he has distinguished.[4] This means that both must learn to recognize a set of grammatical elements, primarily particular phonemes (or phonemic distinctions) and secondarily vocabulary (morphemes, words), in respect to which they speak and perceive utterances. It is this that makes the transmission of an utterance a repetition, whereas an attempt to redo or transmit something whose elements are continuous or not preset is an imitation. It is of course well known that all languages have to be learned; and that when a speaker says something which he has heard he is repeating it, not imitating it (whereas he would be imitating in the case of a yell or other nonlinguistic sound).[5]

---

[2] Of course, complex thought occurs only with language; but it is unclear whether this is related to the discreteness of language, or purely to the complex network of grammatical relations which is available in language.

[3] The centrality of the property of transmissibility is not, of course, directly observable. It is introduced here as an explanatory connection for the observable facts that linguistic elements are discrete, preset, and arbitrary.

[4] This is the familiar principle of reforming pulse signals, as in a digital computer.

[5] For a discussion of repetition vs. imitation, see Kurt Goldstein, Language and language disturbances (1948).

The fact that the grammatical elements must be preset makes possible what has been called the arbitrary character of linguistic signals, i.e., the fact that the sounds out of which words are composed do not suggest the meanings of the words.[6] A natural representational relation between word sounds and word meanings would be useful only if it could be varied, guessed at, and in general if it eliminated the need for the language users to have to learn a fixed and jointly accepted—i.e., preset—stock of elements. But the fixed stock of elements is precisely what is required for transmissibility without error compounding.

Furthermore, if the linguistic signals were not arbitrary, but inherently related to their sound composition or their meanings, they could not, in many respects, be identically preset for all participants. For if the identity and meanings of sounds were not learned, but left for each speaker and hearer to judge on the basis of his individual experience, the speaker and hearer would not necessarily be in agreement, since experience varies in detail for different individuals.

And, most likely, the "meaningful sound" elements would then not be discrete, for there is no known way of arranging and enumerating the fluid world of meanings and feelings in one person's mind, and the differences in respect to this between each person and all others, in such a way as to permit a mapping from stated meaning entities onto stated discrete morphemes and words. Instead, one learns certain sound sequences (morphemes) and one then judges, from the range of combinations in which each morpheme appears, what can be the meaning range associated with each.[7] The discrete and preset character of linguistic signals therefore

---

[6] While certain basic properties are universal for all languages (Chapter 8), there are also great differences as between languages, above all in respect to the sounds used in each language for words which have approximately the same meanings in each language. These differences are not simply due to different meanings expressed in the several languages. The differences are just as great in those items for which the several languages have much the same meanings (numbers, family relationships, basic human emotions, and parts of body) as in those items where the meanings expressed in one language presumably differ from those in others.

[7] The arbitrary character of morphemes in respect to their component phonemes (letters) is made even clearer by the fact that differences between meanings of morphemes can be related to differences in the neighborhoods in which those morphemes occur in discourses, rather than to differences in their sounds. If a word W, for example, has a different meaning in occurrence A (i.e., in neighborhood A) than it has in occurrence B, it is not because W has a different inherent semantic property in A than in B, but because the neighborhood A of W differs correspondingly from neighborhood B of W. And if W has different meaning than word Y, it is not because the phonemes of W differ in meaning from those of Y, but because the range of neighborhoods in which W occurs differs correspondingly from the range of neighborhoods in which Y occurs. In this way, structural linguistics can avoid attributing inherent properties to its elements.

leaves them arbitrary,[8] and indeed requires that there be no general correlation between sound and meaning.

The fact that the grammatical elements (characteristic sounds, and the collecting of these to form morphemes or words) are arbitrary symbols, which are related to meanings only by conventions, makes these elements available for treatment as mathematical objects; for anything which we establish about language as composed of these elements will hold for any other set of elements (e.g., letters) onto which the original set can be isomorphically mapped. Furthermore, the discreteness and fixedness of grammatical elements indicate certain mathematical possibilities as against others.

2.2. Combinations of elements are linear, denumerable

In all linguistic material, the entities (or at least their heads, i.e., their initial segments) can be linearly ordered. Each discourse is a sequence of phonemes. More specifically, each morpheme is a sequence of phonemes, each word a sequence of morphemes, each sentence a sequence of words, and each discourse a sequence of sentences. In certain cases, physical segments of two elements A, B occur at the same time, or intermixed (e.g., B between the beginning and end of A); but in all such cases it is possible to set up criteria that are not *ad hoc* and that determine a unique linear ordering. Thus, in the exceptional case of two phonemes which are pronounced simultaneously, we can write the phonemes in a sequence appropriate to the otherwise established properties of the grammatical structure[9]; in the case of fixed grammatical intonations extending over a sentence (e.g., question) or morpheme (e.g., emphasis) we can mark the intonation at the end (or beginning) of the phoneme sequence over which the intonation extends; in the case of discontinuous morphemes (e.g., when a feminine suffix after a noun is repeated after the noun's adjectives, 6.1, end), we locate the morpheme at the point where it is free to occur or not to occur (i.e., after the noun) and consider the other segments of the morpheme, which depend upon the main segment ("agree" with it), as being deter-

---

[8] There are, of course, determining factors for the signals—the kind and amount of acoustic difference which people can pronounce or hear, and historical sources for the sounds and words of a language. However, any set of sounds or words can be replaced by another equinumerous and similarly structured set without affecting the further grammatical rules which apply to this set. And the individual sound composition of words is not used in any important way in any grammatical rules about the words.

[9] This is the case when a double phonemic distinction (see 3.1) occurs at one point in the phonemic representation of an utterance. Such a double distinction may be represented by two phonemes which are written successively, in accordance with the successions of similar phonemes in similar neighborhoods.

mined by the main segment plus its structural neighborhood (in this case the adjective).[10]

In certain respects, it turns out that the sequence of phonemes can be more precisely represented by two (or more) sequences s_1, s_2 of elements (vari-lengthed components of phonemes). Sequence s_2 occurs simultaneously with sequence s_1 and the length of each element in s_2 is an integral multiple of the lengths of elements in s_1. The successive phonemes of a sentence are the resultant of the simultaneous portions on both sequences. The reasons and methods for such a representation are given in 6.1; but in any case the representation of discourses and sentences remains linear, even if two or more simultaneous sequences of elements are used to represent the phonemic sequence.

The linearity of physical elements is not to be confused with linear orderings in mathematically defined sets for the description of language. In the latter case, we describe language in terms of a set which is closed under some operation, and might ask whether the set is linearly ordered. This turns out to be not quite the case, although we can come close to this result (7.5). In the former case, of the physical elements, we are merely saying that language events are included within the set A of linear orderings of words (or morphemes, or phonemes). The set A that is closed under the word-sequence operation is not natural language, and there is no direct way of characterizing what subset of A a natural language is. But it remains true that in each language event the phonemes, words, sentences, are linearly ordered (at least in respect to their heads).

The fact that discourses and sentences are sequences of discrete elements can be used in considering their number. The set of arbitrary grammatical elements, including sound elements, vocabulary, rules of classification, and rules of combination, must be finite, or recursive with finite generators; for otherwise it could not be preset discretely in the finite speaker and hearer. Hence the grammar of a language (the metalanguage for the structure, 5.4) is finite, in at least one of its forms. Sentences are always finitely long; we can never say whether a particular word sequence is a sentence or not until it is ended, for otherwise something might still be included in the sequence that would violate the regularities for sentencehood. But there is no upper bound to the possible length of sentences, since one can always add some clause or repeat a word, such as *very*. Hence the set of sentences, as sequences of elements in a finite discrete set, is denumerably infinite, even

---

[10] It may be mentioned that such devices are not in general available for the elements constructed in the course of a mathematical type of analysis, as will be seen in 3.6 and 4.4. In a string decomposition of a sentence, the linear ordering of string entries does not uniquely characterize a decomposition; and in a transformational decomposition of a sentence, the transformations are in general only partially ordered.

though it will be seen below that the matter is complicated by the fact that the set of sentences is not well-defined and is not even a proper part of the set of word sequences.[11]

Finiteness entails various properties of grammar, including not only a finite number of operators but also finite generators for all the arguments of each operator,[12] and including the fact that idioms and other exceptions to grammatical rules can be described as extensions of regular operations of the grammar.[13]

2.3. *Not all combinations of elements constitute a discourse*

In every language not all the finite sequences of the phonemes of that language occur as sentences or discourses. The fact that not all combinations occur makes it possible to define larger elements (e.g., morphemes) as restrictions on the combinations of smaller elements (phonemes). This redundancy is essential to natural language, since one cannot always have recourse to a prior language in order to identify the larger elements of a natural language; this is clearly the case for an infant learning its first language. And many or most words cannot be adequately identified by nonlinguistic activity such as acting out or pointing; this applies to some extent to the grammatical words in all language learning. The words and grammatical forms may be taught, in the sense of being singled out, or they may be noticed and finally recognized by the child or other language-learner; but in either case, the possibility of distinguishing the elements requires that not all combinations of elements occur. Consider morphemes, each of which is some sequence of phonemes. If each sequence of phonemes (say, up to some length) constituted a morpheme, and each sequence of morphemes constituted a sentence, there would be no way of identifying the morphemes (e.g., as to meaning), or even of knowing where lay their boundaries within a sentence, except by reference to another language, whether metalanguage or translation. Such reference to another

---

[11] The latter is due to the fact that there are sentences which contain sound sequences that are not words: Any sound can be the subject of a sentence of the form *X is a sound*, *X is his name*, *X_1 and X_2 are different sounds even though we cannot hear the difference* (5.4), etc. The set of objects that occupies the positions of *X* here, and so the set of sentences of the above forms, is not discretely differentiated (aside from the limits of discrimination of hearing and perception) and not necessarily denumerable.

[12] An example of a grammatical fact which satisfies the finiteness requirement is the fact that the resultant of a transformation is, except in a small number of cases, not a new sentence form but one which is similar to some otherwise existing sentence form. Thus we do not have an unbounded set of new resultant sentence forms which would have to be defined as operands for further transformations.

[13] The properties listed in 2.1–2 make it possible to plan orderly search for the discrete, preset elements, in linear combinations, for each particular language.

language, and hence complete utilization of all combinations, is possible for a simple (not error-correcting) code; but not for natural language, which must therefore contain considerable redundancy as to phoneme and morpheme sequences. The fact that certain phoneme sequences, or morpheme sequences, occur, while others do not, makes it possible for the hearer to recognize the boundaries of morphemes within a sentence.[14] Even the meaning of a morpheme A relative to the other morphemes of the sentence can be (and is) determined from the set of morphemes that A occurs within a sentence.[15]

Furthermore, in every language, not all the finite sequences of the morphemes of that language occur as acceptable sentences or discourses. In all languages, the departures from complete utilization of morpheme sequences are such as to permit the setting up of morpheme classes. That is, it is always possible to collect morphemes into classes in such a way that the sentences or discourses of the language are far less redundant as a set of sequences of these classes than they are as a set of sequences of morphemes. The classes of morphemes are (or can be) defined in each language solely by this criterion. Even then, by no means all sequences of the morpheme classes occur as sentences. Here again it turns out that we can define particular sequences of morpheme classes (3.5, 4.2) and then collect these sequences into classes and then state what combinations of these classes of sequences constitute sentences; the restrictions on combinations of these new classes of sequences are far less than the restrictions on combinations of morpheme classes. This whole process may repeat once or twice before we arrive at entities having the least restrictions on combination in constituting the sentences of the language.

It is thus seen that in every language not only is there redundancy in respect to the sequence of ultimate elements (phonemes), but also this redundancy is composed of a system of contributory redundancies, each in terms of intermediate elements. It will be seen that each of these contributions to the total redundancy has meaning: the meaning of entities, and the meaning of grammatical relations among them, is related to the restriction on combinations of these entities relative to other entities.[16] The complex

---

[14] As is done in 3.2.

[15] More exactly, in an elementary sentence and in operators.

[16] This fits with the fact that phonemes, which are not defined on the basis of redundancy of some other entities, do not have meaning. The fact that particular kinds and amounts of redundancy are essential parts of language structure makes it important that a description of language should not add its own redundancy to the picture. A theory of language should not contain elements of wide combinability and then specify which combinations are language. It should contain elements of just such combinability as appears in the language itself.

system of redundancies is thus necessary in natural language in order to provide a complex system of meanings.

The redundancy mentioned here refers not to frequency of occurrence of certain sequences, but to whether certain sequences occur at all as accepted sentences. One might, of course, argue that for every word or phoneme sequence A there is some environment B in which A can occur in an acceptable sentence, e.g., in a nonsense sentence or in a sentence about arbitrary phoneme sequences. However, in that case the combination AB occurs acceptably, while the combination of A with non-B does not, so that it remains that some combinations of elements do not occur as acceptable discourses.

This limitation on combinations raises the question of how to distinguish the sequences (of phonemes, morphemes, etc.) which occur as sentences of the language (or as distinguished parts of discourses in the language) in contrast to sequences which do not occur as sentences. This is the general problem of structural linguistics.[17] It is a problem of finding regularities in those sequences of elements which constitute sentences (or discourses) as contrasted with those which do not. In the denumerably infinite set of word sequences which are discourses, or distinguished discourse segments, such regularities must exist, if the elementary grammatical combinations and operations which the discourses exhibit are finite or recursive. So much so that, as will be seen (4.2.4), even the grammatical exceptions are only extensions of grammatical regularities.

The regularities that are found are not in terms of transitions between the successive elements of a sentence. That transitions among the directly observable successive segments of a sentence do not suffice to characterize sentencehood may be seen for example in the fact that certain sequences of phonemes, and of morphemes or words, are ambiguous. If we consider a sequence such as *It was sent by the senator from Ohio*, we see that it has two sharply distinct meanings (he sent it from Ohio, and the senator is from Ohio) even though the individual words have the same meaning in each case.[18] The difference in meaning has to be due to something other than the mere sequencing of the words or phonemes—to some categorization or

---

[17] Indeed, it is the only general problem (about the whole language) which can be formulated in the terms of structural linguistics (phonemes, morphemes, sentences, etc.), since other problems involve external concepts such as the circumstances in which something is said.

[18] One might argue that the difference in meaning is due to different grammatical classification of the words (e.g., *from Ohio* as adjunct of verb or of noun); but this would be an example of precisely what is being argued here.

grouping of the words, or (equivalently) some choice of forms (classifications, well-formedness) for the sentence.

At this point it is necessary to state the empirical fact that in each language there are periodicities of combination of word classes (or word-class sequences) which are called the sentence structures of that language. Sentences are not directly observable. Each language is a set of discourses. However, it is found in each language that every discourse can be segmented into sections in such a way that each section is a case of one of certain well-formed sequences of word classes or of word-class sequences (or, in exceptional cases, is an initial segment of such a well-formed sequence). These sections are called sentences, and the well-formed sequence is called a sentence structure or sentence form. Each language has only a few sentence structures.

The main activity of structural linguistics has been to define collections (classifications) of elements, or of sequences (of classes[19]) of elements, in such a way that characterizable sequences of the classes are a sentence structure, whereas other sequences of them (or of anything else) are not. Morphemes and words are thus classified into classes of noun (N), verb (V), conjunction (C), preposition (P), etc., and sequences of such classes are classified into subject of sentence, or adjunct of noun, etc. In such terms we can say, for example, that no VCC (e.g., *went but or*) is a sentence whereas NV (*Time flies*) is a sentence. But we cannot say that every NV is a sentence, e.g., *people attributed* is not. It is thus necessary to recognize subclasses within the classes: V_0 for verbs which have no object following (e.g., *exists*), V_{npn} for verbs which have following NPN as object[20] (e.g., *attributed the Jena to Beethoven*). These are subclasses of V in that both occupy the position after N in a certain set of sentences, and in that certain other grammatical properties (of V) are common to both.

But this is not all. We cannot say that every word sequence in the subclasses NV_0 or $NV_{npn}NPN$ occurs as a sentence, or occurs with equal naturalness: *The bird flew* is natural; *The man flew* is more natural now than formerly; *The stone flew* (*through the air*) is natural (although some might consider it metaphor or imprecise); *The house flew through the air* less so; *The mountain flew through the air* much less; *The earth flew through the air* virtually impossible. It is not solely a question of meaning, because there are many combinations where meaning is not decisive, e.g., is *His*

---

[19] The term "class" will be used here for collections of morphemes or words, extensionally or intensionally defined.

[20] "As object" means that there is a form, i.e., an expected well-formedness for the sentence, requiring NPN after V_{npn}, zero after V_0, etc.

words flew through the air a natural sentence?[21] It is also not a matter of frequency or likelihood of occurrence, for *The airplane flew* was perfectly natural the first time it was said, and *The carborundum flew through the air* is natural (in the weakly metaphorical sense of *flew*) even if it has never been said. One might wish to define a subclass $\{N_i\}$ (called co-occurrents, or selection) of N for each member V_i of V, such that V_i occurs with each one of $\{N_i\}$ in the NV sentence structure; but this would be a different sense of subclass than that used for V_0, V_{npn}, and its membership would be changeably graded in naturalness.

Any attempt to distinguish the word sequences which are sentences from those which are not has thus to satisfy the condition that the boundary between the two is not sharp. There are many word sequences which are acceptable as sentences only with special responses, as metaphors, jokes, etc. (e.g., *He is married to his work*, *The words fell on deaf ears*). There are others for which people differ, or cannot decide, as to whether they are sentences. There are cases of two different forms being used for what is intended as one sentence (*A number of people is here*, and *A number of people are here*). There are some sentence forms which are, at a given time, productive, i.e., are used for some but not all members of a subclass but can be extended as neologisms (e.g., *Charcoal erases easily*, *Prefabs build easily*; *American money spends easily?*). All this shows that there are a great many word sequences which are sentences only marginally, and marginally in different ways.

The term "set" is therefore used for sentences only in the unusual sense of a collection in which membership is a graded property, in the manner discussed in 4.1.1. As will be seen (4.1.5), definite and ungraded entities can nevertheless be constructed on this basis. To have a well-defined sense of "set," we can specify a cut-off point in the acceptability required for membership.

In addition there are many idiomatic forms which apply particular grammatical relations to words outside the domain of these relations (e.g., *He met his doom*), and many irregular forms which have unique grammatical operations on unique domains (e.g., peculiar syntax as in *He made a great to-do about it*; morpheme alternants as in *I am, you are,*

---

[21] Of course, we can say that the meaning of *flew* is or is not extended to *words*, but this begs the question, for it defines the meaning criterion by whether the combination occurs in a natural sentence. The meaning of a word is a factor in determining what other words it occurs with naturally (in a given sentential subclass sequence); conversely we can judge the meaning difference between two words in a subclass from the difference in the list of words with which they occur naturally in a given sentential subclass sequence. Cf. footnote 7.

he is). All such individual facts and small-domain operations are formulated to the greatest extent possible in a form that fits them into the main classes and operations of the grammar, as extensions of these. This means that any characterization of sentences will have to provide for a number of operations that differ at least in detail from the others and that apply to only one or a few words.

The characterizing of sentences is thus far from simple. But it will be seen that it can be carried out precisely and in a reasonable way, to any order of detail that is desired.

2.4. *Operations are contiguous*

Talk or writing is not carried out with respect to some measured space. The only distance between any two words of a sentence is the sequence of other words between them. There is nothing in language corresponding to the bars in music, which make it possible, for example, to distinguish rests of different time-lengths.[22] Hence, the only elementary relation between two words in a word sequence is that of being next neighbors.[23] Any well-formedness for sentence structures must therefore require a contiguous sequence of objects, the only property that makes this sequence a format of the grammar being that the objects are not arbitrary words but words of particular classes (or particular classes of words).[24] But the sequence has to be contiguous; it cannot be spread out with spaces in between, because there is no way of identifying or measuring the spaces.[25]

By the same token, the effect of any operation that is defined in language structure, i.e., the change or addition which it brings to its operand, must be in or contiguous to its operand. No space or distance is defined between operator and operand. Of course, later operators on the resultant may intervene between the earlier operator and its operand, separating them. In the description of the final sentence such separations (i.e., the embedding

---

[22] Certain intonations, e.g., comma, may have optional pauses at their end; but these are not like rests which occupy part of a bar: they are not durations which can also be occupied by words.

[23] Sentences, of course, will admit further relations, due to well-formedness boundaries or operators.

[24] I.e., the format is itself only a sequence, but a sequence of particular classes, with a specified beginning and end.

[25] We can specify that between two segments a, b of a well-formed sequence certain other sequences X may appear. However this does not mean that a distance is defined between a, b; it simply means a family of forms: ab and aXb. Furthermore, in most cases, if such an apparent distance can be defined between a and b, it can be of unbounded length: in most cases there is no upper bound to the number of X's that can appear between a and b.

of later operators) can be recognized. But in defining the action of the earlier operator on its operand this separation cannot be identified; the separation can only have been due to a later event.

If follows that if language can have a constructive grammar, then for language there must be available some characterization of its sentences which is based on purely contiguous relations. The contiguity of the successive words is related to this situation, but does not satisfy this requirement, because a sentence characterization cannot be made directly in terms of the successive words in the set of all word sequences. The sentence characterization will have to define well-formed subsequences or operators which will determine the word sequences that constitute sentences; but these subsequences or operators will have to operate contiguously.

2.5. The metalanguage is in the language

Every natural language must contain its own metalanguage, i.e., the set of sentences which talk about any part of the language, including the whole grammar of the language. Otherwise, one could not speak in a language about that language itself; this would conflict with the observation that in any language one can speak about any subject, including the language and its sentences, provided that required terms are added to the vocabulary. Furthermore, there would then be an infinite regress of languages, each talking about the one below it. Observably, the grammar which describes the sentences of a language can be stated in sentences of the same language. This has obviously important effects, including the possibility of inserting metalanguage statements into the very sentences about which the metalanguage statement was speaking (5.7). At least one form of the complete grammar of a language is finite (2.2).

The inclusion of the metalanguage in the language is facilitated by the existence in natural languages of certain distinguished schemata of sentences. One is the classificatory sentence type, whose simplest form in English is N is N_{cl} (where N_{cl} is a classifier noun for the first N: e.g., *Man is a mammal*). A subtype which is particularly relevant to the metalanguage is X is a *word* (X ranging over all words, or over any phoneme or sound sequence) and the like.

Any adequate description of language has to provide for the metalanguage as a not immediately distinguishable part of the language (5.4).

2.6. Language changes

Natural languages are open: new sentences and discourses can be said in them. Furthermore, natural languages change in time. As far as we can see, they do so without at any point in time failing to have a grammatical structure.

At any moment in the history of a language, it is possible to make as complete a grammar of the language as we wish. No item of the language need be left out as undescribable: any item which is not a case of existing rules of the grammar can be fit in (as a special case under special conditions) to some existing rule in respect to which it can be described.[26] That is, we can describe a language at any time t_1 in the usual terms of structural linguistics (Chapter 3), giving as complete a description as we wish and necessarily including a large number of individual facts (operations whose domains are individual words). At any sufficiently distant time t_2 we can do this again, and in all likelihood there will be some difference in the two descriptions, in respect to some items X, due to changes in the language in the intervening period.

Since t_1 and t_2 are arbitrary, and since there are no discontinuous points in a language history (although there may be such in the description of language history), the description of X used at t_1 must be valid up to some period t_i, $t_1 < t_i < t_2$, and the description of X used at t_2 must be valid from t_i and on, without there being a discontinuity at t_i. We conclude that during the period t_i the grammatical items X must have been describable in two different ways. Before t_i the item might be described in one way, to fit it into certain features of the grammar (2.3 end). After t_i it will have changed sufficiently so as to require a different description, fitting it into some other features of the grammar. At t_i both descriptions must have been possible, i.e., at t_i the amount of change in X must have been sufficient to make X fit the t_2 features, but not so great as to make X no longer fit the t_1 features.[27] The situation at t_i is indeed often observable in detailed grammars, e.g., in the case of transformations which are in progress. Thus, in *He identified it by the method of paper chromatography*, we have two sentences connected by *wh* (4.2.2.4): *He identified it by a method* and *The method was of paper chromatography*. In *He identified it by the means of paper chromatography*, we can attempt a similar analysis into *He identified it by the means* and *The means were of paper chromatography*.

---

[26] The possibility of this is not only clear from the existing grammars of languages, but also follows from the methods of grammar construction as seen in Chapter 3.

[27] It does not matter here whether an individual changes his speech over the period t_i, or whether items X are used differently by people who began to speak after t_i as against those who began to speak before t_i. Both situations presumably occur. All that is relevant here is that there is a time t_i during which both a t_1 description and a t_2 description of X fit the grammatical relations that are adequate for describing the speaker–hearer activities within the language community.

This is, however, not very acceptable, and a more acceptable analysis would be to take *by the means of* as a new preposition, similar to *by* itself. The latter analysis is already inescapable in *He identified it by means of paper chromatography*, since ∄ *He identified it by means* (for the ∄, read: *He identified it by means* does not exist, or is not acceptable, in the language).

We can satisfy the situation above somewhat as follows: A grammar contains certain objects (sound distinctions, morpheme subclasses, etc.) and operations on particular domains of these. We say that before t_i, X was outside a domain D of an operation f. After t_i, X was in D under f. During t_i, either X was in some other domain D' on which f operated, with D' being so close to D (relative to the other differences among domains in the grammar) that one could consider D' as part of D or not; or else X was only marginally operated on by f and hence only marginally in D. In either case, X was an exception during t_i, i.e., it did not fit automatically into the rules of the rest of the grammar; but also, as its further course after t_i shows, it could be considered an extension of an operation and domain which were defined before t_i.

The possibility of a structural static grammar at time slices of a changing system is due to the fact that there are similarities and other relations among the various domains named in a grammar and among the various operations named in the grammar. In terms of these higher level systemic relations, the exceptional domains of an operation can be shown to be extensions of one of its regular domains.

This means that a description of language has to provide for the existence of items which don't quite fit into the rules for the rest of the language, but can nevertheless be related to those rules as extensions of their domain or small modifications of their operation.

We have reviewed the main relevant properties of language, including various facts related to these, such as the existence of ambiguous sentences. Any theory of language has to satisfy these conditions; and we will see how these properties determine the kind of mathematical formulation that can be made of linguistic objects, and how these properties show up in a reasonable way in the course of this formulation.

3

Sentence forms

3.0. Characterization of sentences

Given the properties of language noted in Chapter 2, it follows that we should be able to define discrete elements, and should then be able to describe language as certain well-formed sequences of classes of them. We begin with an experimental method for establishing the ultimate discrete elements, the phonemic distinctions, for each language separately (3.1). A recurrent stochastic process on these elements then distinguishes words (3.2), and another and different recurrent stochastic process on words distinguishes sentences (3.6). The latter process can also be stated in the form of an axiomatic theory which, given the word list of a language and a set of axiomatic sequences, obtains the sentences (more precisely, the sentence structures) of the language (3.5); and it can be stated in the form of a simple cycling automaton which does equivalent work (3.7).

This chapter thus provides a model, and in all essential respects a construction, in answer to what has been seen (2.3) to be the central problem of grammar, namely, characterizing those sequences of sounds, or of words, which constitute sentences or discourses, as against those which do not. In Chapter 4 a system of transformational relations among sentences is defined; and as a byproduct of this we obtain a somewhat different characterization of sentences, descriptive rather than constructive (end of 4.3.2).

The set of steps sufficient to characterize sentences shows what kind of science structural linguistics is. Once the ultimate elements have been established, a natural language can be defined in mathematical terms as that set of sequences of these elements which is characterized by certain peculiar stochastic processes or by a certain axiomatic theory. However, the determination of the elements is as important and as characteristic for the science as the operations upon the elements. As will be seen in 3.1, the elements are determined by speakers' identical recognition of a relation of "repetition" between utterances. The determination of the elements is therefore not, so far, in physical terms. It has not so far proved possible to characterize and predict the phonemic distinctions by acoustic differences, although given the phonemes we can show the differences between them in sound-wave properties. The elements also are not semantic in the sense that any consideration of meaning or variation in nuance can affect the phonemic composition of a word. Phonemes cannot be determined by investigations based purely on meaning, although most morphemes, as sequences of phonemes, are associated with particular meanings. The phonemes also cannot be defined in terms of human functioning or psychological responses, although recognition of the elements and their use is part of the normal functioning of the individual, and disorders of language (as in aphasia) are part of disorders of functioning. Finally, the determination of phonemes is not a social item, at least not in the sense in which "social" is used so often today as a euphemism for institutions and control; no social decision can change the phonemic distinctions. The repetition relation is a learned pattern of behavior common to all the people who interact in speaking or writing. In a model of language, this relation between events (between utterances, or between their segments) can be replaced by the relation between event and class of events, or between "token" and "type"; a is a repetition of b if and only if there is a class of events (a "type") x such that a belongs to (is an occurrence of) x and b belongs to (is an occurrence of) x, where x is determined by the pair test (3.1).

3.1. Phonemes by pair test

In keeping with the first paragraph of 2.1, we first establish a method for determining the discrete sound elements of a language as distinctions (cuts) among subsets of sound events in that language.

Two utterances, i.e., occurrences of speaking or writing, may contain cases (occurrences) of the same linguistic elements. The first question is how to determine a set of elements, not further decomposable into smaller linguistic entities, such that we can say definitely for any two segments a, b, of an utterance that $a = b$ (i.e., a is an occurrence of the same element as is b) or that $a \neq b$. This can be achieved with reasonably good experimental results by the pair test, in which we select two utterances, A and B, preferably very short, with one speaker of the language pronouncing a number of his repetitions of each utterance, randomly intermixed, while another speaker of the language indicates which of these pronunciations are repetitions of each other. In most cases the hearer will guess correctly what the speaker considered repetitions either in close to 100% of the cases or else in c. 50%. In all cases, if the hearer's list of what were the repetitions has a close correlation with the speaker's list, then the two given sets of repetitions are phonemically distinct; i.e., there is a phonemic distinction between them: $A \neq B$. (This would be the case if the two repeated utterances were, for example, *heart* and *hearth*.) Otherwise no phonemic distinction can be asserted between the two given utterances. (This would be the case if the two repeated utterances were *heart* and *hart*.) In some cases the results of the pair test are problematic, and in some cases the decision as to whether a phonemic distinction exists, and of what kind, is adjusted on the grounds of later grammatical considerations. But the direct results of the pair test furnish a starting point, a first approximation to a set of ultimate elements adequate for characterization of language.

Utterances are phonetic sequences of sounds or symbols. One can rerun the pair test on tape recordings which are cut so as to reveal what segment of the utterances sufficed for the distinction which the test had established. Therefore, given $A \neq B$, we can locate the phonemic distinction δ between A and B at the phonetically appropriate point p in A and q in B, and say that A contains a segment x at p and B a segment y at q, the difference between x and y being δ.

If A at p also differs from utterance C at some point, by the pair test, x in A also differs correspondingly from the corresponding segment z at the appropriate point of C. We thus move from distinctions between segments of utterances to distinct segments of utterances. Then t of *hart* represents the distinction of *hart* (or *heart*) from *hard*, *harp*, etc., while d of *hard* represents the distinction of *hard* from *hart*, *harp*, etc.

If we compare the results on various pairings of a few short utterances, we can describe some utterances as being distinct from others at more than one point. Thus we can make a segmentation such that *hart* $\neq$ *hearth* in the last segment, *hart* $\neq$ *dart* in the first segment, *heart* $\neq$ *hurt* and *dart* $\neq$ *dirt* in the second segment, and *heart* $\neq$ *dirt* in the first two segments, and *heart* $\neq$ *dearth* in the first, second, and last segments (despite the spelling). For each utterance, the minimal number of distinctions necessary to distinguish it from every other utterance, in terms of pair tests, determines the number (and location) of segments which express these distinctions. These are the phonemic segments of the utterance.

The number of different phonemic segments can then be reduced by collecting segments x_1, x_2, into one class of segments if all the neighborhoods (i.e., all phonemic segments immediately preceding and following, up to a reasonable distance) of all occurrences of x_1 differ in phonemic composition from all neighborhoods of x_2. A single phonemic symbol X can obviously be used for both x_1 and x_2, for we can always tell from the phonemic neighborhood whether X represents x_1 or x_2 (which are called variants of X). These classes of phonemic segments, each of whose "variant" members differs in its neighborhood from each other member, are the phonemes of the language, roughly the letters of its alphabet.[1]

---

[1] Phonemic analysis was developed between 1910 and 1935 by Ferdinand de Saussure, Edward Sapir, Leonard Bloomfield, Nikolai Trubetzkoy, and others. There are various practical ways of determining the phonemes of a language. The pair test is a particular experimental device showing that a sufficient basis is the relation of being a repetition.

It is clear that every utterance (with its repetitions) has a unique position in the network of phonemic differences among utterances. Not all combinations of phonemic differences occur, however, and when we are in the process of representing the phonemic distinctions by classes of segments, we may find that some utterances can be represented by more than one not otherwise occurring combination of classes of segments. Naturally, we avoid this redundancy, either by stating what types of combination of phonemes do not represent utterances, or preferably by changing the definition of phonemes so that the nonoccurring combinations fall outside the definition. For a suitable definition of phonemes, then, we can say that each utterance has only one phonemic representation, i.e., only one spelling.

Problems often arise in deciding what phonemic segments should be collected into one phoneme. This is done in such a way as to permit the maximum freedom of combination for each phoneme, and the greatest regularity of combination for all phonemes of one type. In some cases it is even desirable, for these ends, to replace the phonemes by elements having more degrees of freedom (simultaneous combinations, multiple lengths; see 6.1).

In this way, using the fact that utterances are composed of parts and that there is repetition rather than imitation of utterances, we are able to determine segments in the set of repetitions of one utterance which differ from segments in the set of repetitions of another utterance. And we can determine a minimal set of different phonemes (as classes of segments) sufficient to distinguish every two utterances which are not repetitions of each other.[2] It follows from the way in which the phonemic representation was constructed that it is unique: utterances A and B are repetitions of each other if and only if they are represented by the same sequence of phonemes.

The phonemic symbols, of which an utterance is composed, are only secondarily classes of segments of utterances. Primarily they are symbols for the distinction of the given utterance from all other utterances from which the given utterance is distinct (by the pair test) at the location of the symbols in question.

---

[2] The relation of being a repetition is not coextensive with that of being grammatically identical. Two homonymous, i.e., grammatically ambiguous, utterences are repetitions of each other: e.g., *We saw them every day* (with our eyes); *We saw them every day* (with a saw). The whole problem of ambiguity is how to recognize those grammatically distinct utterances which are repetitions of each other, i.e., which cannot be distinguished by a hearer except with the aid of further material. Phonemic distinction, as a representation of a hearer's linguistic ability to distinguish, is thus an independent property of utterances, and is a stable datum; it does not vary significantly from test to test or from speaker to speaker, within what is called a language community.

3.2.   Recurrent dependence process: Morphemes by phoneme neighbors

It follows from 2.3 that there is not simply a single redundancy governing the combinations of the sound elements, but several redundancies, as to the combinations of sound elements in forming certain larger entities, and as to the combinations of these entities in forming yet larger ones, and so on. A method for determining the first redundancy, and hence the next larger entities, follows.

Having now a unique representation of each utterance as a sequence of phonemes, we have next the problem of stating in a regular way which sequences constitute sentences of the language. It is impossible to describe all the phoneme sequences of the language in a nonhereditary way[3]; e.g., to say that for each finite sequence of phonemes, no matter what precedes it, certain phonemes may follow and others may not, in sentences of the language. In 3.5 and in Chapter 4, it will be seen that there are unbounded possibilities of embedding segments of sentences inside of other segments, so that, aside from certain nonoccurring short sequences of phonemes, we cannot say that a particular sequence of phonemes can never be followed by a particular phoneme.[4]

It is possible, however, to find a particular kind of regularity in those phoneme sequences which can represent sentences,[5] or initial parts of sentences, as contrasted with those phoneme sequences which do not. In the set Q of phoneme sequences which represent sentences, each initial subsequence q of particular phonemes has a certain number v of different phonemes following next after it in all members of Q in which q occurs. In each member of Q, containing n phonemes, the follower-variety v for the first m phonemes is found to have a sawtooth gradual fall and sharp rise as m varies from 1 to n. This regularity is not found in the set of phoneme sequences which do not represent initial parts of sentences.

Specifically: we take an arbitrary phoneme sequence $ab \ldots n$, and ask how many different phonemes follow the initial sequence a in all members of Q which begin with a; this gives $v(a)$. We then ask how many different

---

[3] "Hereditary" indicates here the dependence of the nth phoneme, p_n, of a sequence on the whole initial sequence $p_1, \ldots, p_{n-1}$.

[4] It is even more irrelevant to state probabilities of transition from a phoneme or phoneme sequence to a following phoneme. Some words (possibly ones whose first phoneme is a rare one) occur so rarely, at least in a given neighborhood, that the transition in that neighborhood to the phonemes of the given word has almost zero probability. Yet when it occurs, it is a perfectly good sentence, e.g., the phoneme sequence $d\bar{z}zw$ in *He bathes Zouaves there.*

[5] No matter how odd or marginal these may be as sentences.

phonemes follow the initial sequence *ab* in all members of *Q* that begin with *ab*, obtaining $v(ab)$; and so on up to the followers of the initial sequence *ab* ... *n*. If the numbers *v* decrease gradually up to some point, jump there to a higher value and decrease gradually again, and so on to the end (where the number of followers is high again), then the *ab* ... *n* can represent a sentence. Otherwise it cannot. E.g., if we take a short sentence, written phonemically, we have the pattern shown by Fig. 3.1.

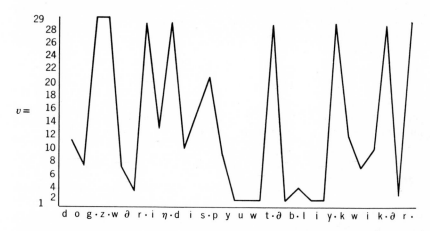

Fig. 3.1. *Dogs were indisputably quicker.* Dots have been inserted between the phonemes, to show where a morphemic segmentation would have been made on syntactic grounds. These dots were, of course, not involved in the test.

If we segment the given utterance at the peaks of the successor count, we obtain segments which appear in many combinations with other similarly determined segments in the other utterances of the language. These segments are the morphemic segments which are the ultimate elements for the rest of the grammar, after phonemics.[6]

We thus have not only a distinguishing property of those phoneme sequences which constitute initial segments of sentences, but also evidence that sentences (differently from nonsentences) contain a certain segmentation into subsequences of phonemes; and we have the boundaries of these morphemic segments.

The peak values of *v* accord more closely with what proves later to be the

---

[6] The peaks result from two facts: that not all phoneme combinations are utilized in making morphemes, and that the number of morphemes in a language is from two to three orders of magnitude larger than the number of phonemes. The boundaries of a morpheme are points of greater freedom for phoneme combination.

morphemic segmentation, and the few cases of gradual rise become sharp, as they should be, if we take v not as the number of phonemic followers to each initial sequence $p_1, \ldots, p_m$ (where p_m is some phoneme a and p_{m-1} is b) but as the ratio of that number to the number of different phonemes that follow the phoneme a (or the sequence of phonemes ba) in all neighborhoods in which a (or ba) occurs.[7] In this ratio of hereditary to Markov chain dependence, we obtain the hereditary effect more sharply; i.e., the restriction on the followers of $p_m = a$ due to that particular p_m being located in a complete initial sequence $p_1, \ldots, p_{m-1}, p_m$, rather than simply in the phonetic neighborhood $b = p_{m-1}$.[8]

Certain peaks in some sentences do not correlate with a morphemic segmentation of that sentence. This happens when the first part of a morpheme in one sentence is homonymous with a whole morpheme which can occur in the same initial neighborhoods. Thus in *He was under ...* there would unavoidably be a peak after *He was un* because of the prefix *un-* which occurs here with many followers (*He was unreserved, He was uninformed*, etc.). Such wrong segmentations can almost always be corrected by carrying out the corresponding count (of predecessors) backward on the same sentence.

Certain other morphemic segmentations cannot come out directly from this method, e.g., in the case of infixed morphemes, and of intercalated morphemes (where morpheme $A_1 A_2$ combines with morpheme $B_1 B_2$ to form a word $A_1 B_1 A_2 B_2$)[9]; or in the case of morphemes which merely change the phonemes of the morphemes to which they are added (e.g., the past tense morpheme which changes *take* to *took*, although after other words it adds an easily segmentable *-ed*). Close study of the results of the next-neighbor counts (including the number of followers of the next neighbor) suggests modifications which make the experimental results accord more closely with the morphemic segmentation that proves useful later. In any case, the segmentation of a sentence into words is given immediately by this method, and the further segmentation of prefixes, suffixes, etc., is given to such a high degree that the rest can be readily obtained by later comparisons of sentences.

---

[7] We may consider the pair ba, rather than just the last phoneme, because many syllabic restrictions depend on pairs of neighboring phonemes. The way in which the phonetic property of phonemes restricts their successor does not depend on the predecessors back to the beginning of the utterance.

[8] In the example above, the two cases in which v rises gradually to the peak, 9–14–20 for *dis* and 11–6–9–28 for *kwik*, would thus be replaced by decreasing ratios. Fewer phonemes can follow a consonant, and in particular a pair of two consonants, than can follow a vowel. Thus 9 is a smaller percentage of the total followers of i than 6 is of the total followers of w.

[9] E.g., word roots and grammatical affixes in some languages.

Results almost as strong are obtained if we start the count afresh after each peak rather than from the beginning of the phoneme sequence. In that case the decreasing number of different successors to each successive phoneme depends only on the history of phoneme selections since the last word peak. If we go through a sentence or discourse or other long sequence of phonemes, the word peaks become recurrent events of the stochastic dependence. An example of the dependences in this recurrent stochastic process[10] is given in the list below of individual words, each tested against all the words in English.[11]

```
        1     1     2     1     1     2    20     5    13    25    ←
   d     i     s  •  t     u     r     b  •  a     n     c     e
→     15    24    24     8     2     2     4     2     1     1
      10.2   9.2   6.4   3.9   1.5    2    1.7    1     1     0


        1     1     4     3    18    15    11    25    ←
   d     i     s  •  e     m  •  b     o     d     y
→     15    24    24    11     5     6     5     2
      10.2   9.2   6.4   2.5    2     2    1.2   0.5


        4     7     1     1     5    24    12    26    ←
   d     i  •  s     u     l     f  •  i     d     e
→     15    24    24     5     2     4     2     1
      10.2   9.2   6.4   1.4   2.5   1.5    1     0


        1     4     2     2     9    19    17    25    ←
   d     e  •  f     o     r     m  •  i     t     y
→     15    26     9     5     4     4     3     1


        5     9    24    26    ←
   a     p     p     l     e
→     26    14     7     5
      15    7.7   4.6   4.2
```

[10] The probabilities for each outcome, i.e., each possible successor, are of no interest here, and are not given. The process states only the number of outcomes, at each state, which have any positive probability.

[11] The next-phoneme count was made by a computer into which was fed Webster's unabridged English dictionary, together with various specialized science dictionaries, alphabetized both forward and backward. Unavoidably, ordinary spelling was used here instead of phoneme sequences, but the method remains applicable, although the spelling results may correlate less sharply with morpheme boundaries.

The figures above the word give the number of different phonemes which precede the final sequence to their right: e.g., 5 phonemes precede final *nce*, 20 phonemes precede final *ance*. The first row of figures beneath the word gives the number of different phonemes which follow the initial sequence on their left; the number below each figure gives the average number of followers to each follower. Thus, 8 phonemes follow initial *dist* (in all words beginning *dist*-), and these 8 have an average of 3.9 different phonemes after them in this initial sequence. Correction for the ratio to Markov chain neighbors has not been made here. As before, the morpheme-separating dots were not included in the text; they were inserted here in order to judge the result.

This sequential dependence, among the phonemes of sentences, segments the sentence into parts called morphemes (and this without any appeal to meaning). In what follows, we shall see that stated sequences of these morphemes define the syntactic elements (constituents 3.4, strings 3.5, or elementary sentences and operators 4), and that stated sequences (plus permutations and phoneme changes) of the syntactic elements define the set of all sentences. Phoneme sequences which are not segmentable in this way into morphemes do not satisfy the further rules of combination, and are not sentences.

3.3.   Well-formedness in terms of classes

It was pointed out in 2.3 that the great restrictions as to which morpheme sequences constitute sentences are of such a kind as to make it possible to collect morphemes into classes on the basis of their gross restrictions and then to state far simpler restrictions as to which sequences of classes appear in sentences. We consider here the establishment of these classes and their sequences.

The preceding section shows that the sound sequences which constitute sentences, when specified as phoneme sequences, are segmented into morphemes which occur in various sentences. If we now wish to describe sentences, even quite simple ones, as sequences of morphemes, we find that there are thousands of morphemic elements with thousands of restrictions on their combination. However, it is possible to isolate the problem of these restrictions (and indeed to utilize it later on) by collecting the morphemes or words into classes (and subclasses), and defining variables, each of which ranges over the members of a class or subclass. Certain sequences of these variables are called well-formed: this means that for some values in their domain these sequences occur as acceptable sentences or as structurally defined components of sentences. Well-formed sequences of word classes are called a form. Thus we define N ranging over *water*, *butter,* and other nouns; *t* ranging over *-ed* (past tense), *can*, etc.; *V* ranging over *fall, melt*, and other verbs. Then *NtV* is a well-formed sentence form, because *Butter can melt*, etc., are acceptable sentences, although *Water can melt* may not be.

An essential property of these variables (utilized in 4.1.1 and justified in 7.3) is that in an *n*-variable form not all *n*-tuples of values of the *n* variables occur with equal acceptance or in identical larger neighborhoods.[12] If we let each *n*-class sequence have different acceptability values for different *n*-tuples of its values, we can collect morphemes into classes, in spite of the many restrictions on sequences of morphemes. A well-formed formula in logic or mathematics has a value (T or F, or numerical values, etc.) for each set of values of its variables. A well-formed linguistic form has a value of acceptability for each set of values of its variables; zero acceptability for some *n*-tuples of values, normal for others, intermediate values for others, unstable values for still others.

In the course of collecting morphemes into classes, we often find that morphemic segments which are complementary as to their neighborhoods can conveniently be considered variant forms of the same morpheme. Thus *am* in certain neighborhoods of *I*, *are* with *you* and plurals, *is* elsewhere, can be considered variants of *be* plus present tense. Similarly *knive-* before plural *-s*, and *knife* elsewhere, can be considered variants of one morpheme. We thus recognize morphophonemics, i.e., the changing of the phonemes of a morpheme, depending on its neighboring morphemes. One can even define successive morphophonemic symbols, whose values are the successive phonemes of a morpheme in each of its relevant morphemic neighborhoods; and one can say that the morpheme is a sequence not of phonemes but of these morphophonemes.[13] This is especially convenient if we start with the morphemes as primitively given.

It is also possible to include short sequences of morphemes as members of a morpheme class if the class variable takes these sequences as values in the same way it takes single morphemes. E.g., *building* (in the sense of *house*) would be a member of the same subclass of *N* as is *house, book*, etc.

3.4. *Morpheme neighborhoods as sentence constituents*

The well-formed class sequences are established by some finite set of metalinguistic rules (2.2). If they are to represent the denumerably infinite

---

[12] If a form *AB* is acceptable for values a_i of *A* and b_j of *B*, the relation of selection, or co-occurrence, is said to hold between a_i and b_j.

[13] Thus *knife* is the morphophonemic sequence *NAYF*, where *NAY* = the phonemes *nay*, and *F = v* before the morpheme *-s* plural (but not *'s* possessive) but *F = f* otherwise.

set of sentences (or discourses), they must be combinable by rules of combination, some of which must be recursive. By 2.4, the entities which are connected in a rule of combination must be contiguous. Before presenting a model that satisfies the requirements of contiguity, we note a traditional model for language which does not quite succeed. In this traditional model, X is an expansion of A if whenever AB is a well-formed sequence, so is AXB (or XAB). AX (or XA) is then a substituent of A. The inadequacy of this model is due to the fact that, in various languages, cases could be found of well-formed sequences which were expansions without being contiguous.

Certain sequences of morpheme classes constitute short sentence forms, e.g., the NtV above. However, the unboundedly many longer sentence forms cannot simply be listed. They have to be characterized on the basis of their satisfying some condition, one which presumably the short forms also satisfy. Various such conditions will be discussed in 3.5, 3.6, and Chapter 4.

A whole family of methods has been based on the condition that in many languages the major types of sentences (excluding questions, imperatives, etc.) can be considered as consisting of a subject, a verb, and its object (and various added segments: subordinate clauses, special words such as *moreover*, etc.).[14] A finite grammar can be produced only if it is possible to specify finitely the sequences of morpheme classes that constitute a subject, etc.[15] In many languages, it turns out that there are only a few types of sequence for, say, the subject, and that the main type always contains a particular word class (in the case of the subject, a noun) with often certain neighboring classes (e.g., adjective). If various neighbors were adjoined to the central class, the successive adjoining could be presented in the hierarchical form of an increasing expansion of the central class. The customary form of grammar therefore said that each of these sentence segments was a constituent or phrase consisting of a central word class to which certain neighbors could be added (semantically as modifiers),[16] and which in some cases could be replaced by certain other sequences.

[14] This is similar to the various parsing methods of school grammar, and has been codified by Leonard Bloomfield and his successors as analysis into immediate constituents.

[15] In many languages, such sentence segments as subject can be more simply described as sequences of word classes, where a word class is a regular sequence of one or more morpheme classes plus phonetic features of a word, such as stress. Below, A and S are variables ranging over adjectives and sentences, respectively.

[16] These expansions of central word classes are organized into a hierarchial system of equations, of the forms $A = AC$, $A = BC$, in Z. Harris, From Morpheme to Utterance, *Language* 22 (1946) 161–83. They are codified into a hierarchical system of rewrite rules in Noam Chomsky, *Syntactic Structures* (The Hague 1957).

Thus the subject is a noun phrase, which is defined as N with possibly A, etc., adjoined (e.g., *large books, the book here*); but in some cases the N is replaced by pronouns (e.g., *This*), *the A* (e.g., *The larger*), certain *wh*-clauses (e.g., *What he planted*), *that S* (e.g., *That he was here*), etc. The operation of adjoining can itself be made a special case of replacement, by saying not that A, etc., is adjoined to N, but that N is replaced by AN, etc.

However, in many and perhaps all languages, the replacement is only a secondary effect. Some of the cases of apparent replacement are due to an adjoining which has been followed by zeroing of the central class: *this* from *this N*; *the A* from *the AN* (e.g., *the larger book*), *what he planted* from *the plant* (or: *thing*, etc.) *which he planted*.[17]

In other cases, we find that the replacement segment occurs in neighborhoods where the central word class does not occur. Thus we have *That he was here is a fact*, but the subject of *is a fact* is not generally a noun. It can be shown[18] that N (with its adjoined neighbors) is the subject of certain verbs (or predicates), while segments like *That S* are the subject of others.[19] The *That S* does not therefore replace the N in any sentence or sentence form in which the N had appeared prior to the replacement.[20] Instead of this replacement, then, we have to recognize a family of similarly structured sentence forms[21]: N (which is interpreted as subject) followed by certain classes of V; *That S*[22] (also as subject) followed by other classes of V.

As to the constituents which are formed by adjoining modifiers around the central word class, there is a point of difficulty in defining the whole process of adjunction: Although most adjunctions are in the immediate neighborhood of the central word class, some are not. The adjective of the subject in Latin can occur at various points of the sentence; in English a *wh*- clause (4.2.2.4) can be away from its noun (usually if no other noun intervenes): *Finally the man arrived whom they had all come to meet.*

---

[17] These zeroings are shown to operate, here and elsewhere, in a regular way (4.2.2.6). It is no accident that most pronouns (e.g., in *This fell*) are similar or identical to pro-adjectives (e.g., in *This book fell*).

[18] In 4.2.2, under ϕ_s; details in various issues of Papers on Formal Linguistics.

[19] Many verbs can have both types of subject, but this is due to various transformational developments.

[20] Of course, it is possible to set up a symbol, Σ, for subject, and say that both N and *that S*, etc., replace it in the sense of values replacing a variable.

[21] We would like to say that the overall subject–predicate structure results from some simple process of sentence building, but fail here. In transformational analysis, we can say that it results from the similarity of the set of resultants to the set of operands (4.2.3). Transformational analysis also shows that forms which lack this structure, like the question, are derived from forms which had it.

[22] And *whether S, N's Ving* (as in *John's arriving*), etc.

In describing sentences, one can still say that there is a constituent, even though with noncontiguous parts: the subject above is *man* with adjoined *the* on the left and *whom* ... after the verb on the right.[23] But the difficulty lies in formulating a constructive definition of the sentence. For if we wish to construct the sentence by defining a subject constituent and then next to it a verb (or predicate) constituent, we are unable to specify the subject if it is discontiguous, because we cannot specify the location of the second part (the adjunct at a distance). At least, we cannot specify the location of the distant adjunct until we have placed the verb constituent in respect to the subject; but we cannot place the verb in respect to the subject as a single entity unless the subject has been fully specified.[24]

A similar constructive difficulty arises with respect to all other non-contiguous phenomena such as discontinuous morphemes, and the special case of them known as grammatical agreement (as between subject and verb in *The man walks* but *The men walk*).

The existence of such phenomena does not mean that a constructive definition of sentences is impossible. It only means that we have to formulate the construction of the sentence not on the basis of word classes to which are made adjunctions up to a whole constituent, but on the basis of some larger entity, some elementary sequence of central classes, in respect to which there are no noncontiguous phenomena (2.4).

3.5. *Axiomatic strings for contiguous operation*

Since we have seen (2.4) that a constructive theory of sentences requires that all relations between components be contiguous, and since the constituents in the sense of 3.4 are not always contiguous, we seek now to define the least entities out of which all sentences can be constructed without noncontiguities. To this end, we define for each language certain elementary strings of morpheme classes or word classes which satisfy the following properties: The strings (whether elementary or also nonelementary ones, below) are collected into sets, $s_1, s_2, \ldots, s_m$ in such a way that for each i either the strings of set s_i are a sentence form, or else they can be inserted (in the course of constructing a sentence) into a string of some set s_j at a stated point (interior or boundary) of that string, the result of inserting a string of s_i into a string of s_j being a string (nonelementary) of s_j.

---

[23] And one can specify that it can be at a distance primarily if no noun intervenes.

[24] To the extent that such problems did not arise, it would be possible to define sentence forms as short sequences of morpheme classes (or word classes), each class being expandable by a certain neighborhood of other classes.

We thus have a set of elementary strings, called center strings, which are not inserted into any other strings and are themselves sentence forms. Examples in English are; NtV_0 (V_0: verbs which have no object, e.g., *Crowds may gather*); $NtV_n N$ (V_n: verbs that have object N, e.g., *Crowds ringed the palace*); *That S t be N_s* (e.g., *That he arrived is a fact*); and also the question forms, etc. Strings of other sets are inserted into these and into strings of various other sets. For example, into any string containing N there may be inserted certain strings (called left adjuncts of N) to the left of the N: this includes adjectives, *the*, etc., in a certain order; and there may be inserted other strings (called right adjuncts of N) to the right of the N: this includes *VingN* or *PN*, as in *demanding their rights* or *from the country-side* inserted in *Crowds ringed the palace*. A subset of these right adjuncts may also be inserted at the end of the host string: *People gathered, who demanded their rights*. Another set of strings (called right adjuncts of V) may be inserted to the right of V in any string containing V: this includes *PN* or *D*, as in *in a rush* or *quickly* inserted after *He arrived*. Of course, since the left and right adjuncts of N can be inserted into any string containing N, they can be inserted into the *PN* right adjunct of V: The result of inserting *great* in the verb adjunct *in a rush* is a string of the verb-adjunct set (but no longer an elementary string) *in a great rush*; this string can be inserted wherever *in a rush* can.

In English (and many other languages), the sets of strings required for a description of all sentences are:

Center strings (some 10 partially similar families of forms, not counting the distinctions due to different types of subject or object which a verb takes);

left adjuncts on *P* (prepositions) and *D* (adverbs) (e.g., *almost* as adjunct on *at*, *immediately*), more precisely on strings containing that *P* or *D* but inserted to the left of the *P* or *D*;

left, or right, adjuncts on adjectives (e.g., *very* or *in intention* on *serious*), on nouns, and on verbs, i.e., on strings containing those words, but inserted to the left, or right, of those words;

adjuncts on center strings and on centerlike adjunct strings (e.g., subordinate clauses), entering at the left or right or at interior points of the strings;

right adjuncts (conjunctional) on virtually any string or any segment of one (e.g., *and* ...).

For each language the strings are determined on the basis of contiguity of operation. Consider any restriction, e.g., as to which word subclass occurs with which other word subclass, or which individual word of one class occurs with which individual word of another, or the locating of a grammatical agreement on a pair of classes. Every such restriction occurs between the members of a string (e.g., plural agreement between its N and its V), or between part of a string and an adjunct inserted into that string (e.g., the restriction of a pronoun to its antecedent, or of an adjunct to the particular word, in the host string, to which the adjunct is directed), or finally (and very limitedly) between two related adjuncts of a single string (related either in the sense of being successive adjuncts at the same point of the host, as in the order restriction in *Crowds from the countryside demanding their rights ringed the palace*; or, very rarely, in the sense of being two parallel adjuncts at corresponding points of the host string, as in the zeroing restriction between the two adjuncts in *People who smoke distrust people who don't*).

Within these conditions, there is room for a certain variety. For example, a string $b_1 b_2$ could be inserted at two points of its host $a_1 a_2$ instead of at one, yielding, e.g., $a_1 b_1 a_2 b_2$ instead of the usual $b_1 b_2 a_1 a_2$ or $a_1 b_1 b_2 a_2$ or $a_1 a_2 b_1 b_2$; this happens in the rare intercalation *He and she play violin and piano respectively*. But one adjunct c of a host $a_1 a_2$ could not be inserted into another adjunct $b_1 b_2$ of that host: we would not find $a_1 b_1 c b_2 a_2$ or the like (unless c was an adjunct of $b_1 b_2$). And a part of a string cannot be dependent on an adjunct to that string, or on anything else which is outside the string.[25] We call "strings" those sequences of classes which satisfy these conditions; and what is empirical is that it is possible in each language to find class sequences (and of rather similar types in the various languages) which indeed satisfy these conditions.

A good approximation to the strings of a language can be obtained by successively excising, from each set of what are tentatively judged to be similarly constructed sentences, the smallest parts that can be excised preserving independent sentencehood (i.e., where the residual sentence does not depend on the excised portion and retains certain stated similarities to the original sentence), until no more can be excised: each excised segment is an elementary adjunct string, and the residue is an elementary center string. It follows that the successive word classes within a string are required by each other, while each excisable string is an adjunct,

---

[25] In the case of idioms, where a string member may require a particular adjunct, we have to make the adjunct part of the string. E.g., in *in the nick of time*, we have *nick* only if *of time* follows (differently from *at the end of the day* where *of the day* is not required and is an adjunct); this has to be taken as a single string.

permitted but not required by the residue from which it was excisable.[26] Thus in *Crowds from the whole countryside demanding their rights ringed the palace*, we can first excise *whole* and *their* and then *from the countryside* and *demanding rights*. In constructing the sentence from elementary strings, we first adjoin the two left adjuncts of *N*, *whole* and *their*, to the two elementary right adjuncts of *N*, *from the countryside* and *demanding rights*; and then the two resultant nonelementary right adjuncts of *N*, *from the whole countryside* and *demanding their rights*, are adjoined to the elementary center *Crowds ringed the palace* (to the right of a stated *N* in it), yielding the desired sentence.

The result of applying the criteria mentioned here is that we are able to set up a small number of sets of strings, with from 5 to 20 types of string in each set, which suffice for the construction of all sentences, and which satisfy various additional properties. For example, no further operation on the strings can introduce intercalation of them (beyond what may be given in the original insertion conditions for particular strings, as with *respectively*), because any permutation can only be on the word classes within a string, or in moving a whole string from one point to another in the host string (as in permuting adverbs to new positions). No non-contiguity arises, because whereas an adjunct, e.g., of a noun, may be at a distance from that noun, it can never be at a distance from the elementary string which contains that noun.

With the analysis of 3.5, the central problem of grammar (2.3) approaches solution. We can state which sequences of word classes are sentence forms, although not which sequences of words are sentences.

The various relations among the word classes of a sentence form, including those that can be obtained from transformational analysis, hold among such words of the sentence as have a string relation between them in that sentence. These further relations, such as transformations, can

---

[26] There are, not surprisingly, various problems. The empirical methods make into string members, or into adjuncts, certain segments which we would like to have the other way around, for reasons of the overall properties of string membership and adjuncthood. Thus most strings containing *N* should be defined as containing *TN*, where *T* represents *a*, *the*, numbers, plurals, etc.: $\nexists$ *Book fell;* only $\exists$ *A book fell*, *The book fell*, *One book fell*, *Books fell;* nevertheless we would like to say that the morphemes here listed under *T* are adjuncts. Furthermore, the fact that words can be zeroed under certain circumstances makes an apparent adjunct out of certain material which we would like to consider parts of the string. E.g., the zeroability of many objects would make *books* an adjunct in *He reads books* (because we also have *He reads*). Such cases can be decided only if, in determining string definition, we include the requirement of certain similarities to other string sets.

therefore be stated in string terms. For example, the two meanings (and grammatical analyses) of *These doctors are attending physicians* are given by *Ving* as left adjunct on *physicians*, and alternatively by *VingN* as object in an *N is* Ω string (Ω for the stateable list of string completers, i.e., objects, after the *V*). The two meanings of *He walked and talked quickly* are given by *and he talked quickly* with zeroed *he*,[27] as adjunct to *He walked*, and alternatively by *and he talked quickly* with zeroed *he* and zeroed *quickly*, as adjunct to *He walked quickly*.

However, word dependence among the words of an adjunct and its host is not directly stateable in string terms. Thus *demanding their rights* occurs with *Crowds ringed the palace*, but *rotating around the earth* is most dubious there. All nontransformational grammars, including constituent analysis, are unable to utilize such considerations. Transformational analysis succeeds here, by deriving *VingN* from *NtVN* and deriving noun adjuncts from two sentences with a shared *N*. Then *Crowds demanding their rights ringed the palace* is derived from *Crowds ringed the palace* and *Crowds demanded their rights* (by the *wh-* operator). And *Crowds rotating around the earth ringed the palace* would be obtained only to the extent that *Crowds rotated around the earth* was available.

The other property which string analysis (and in different ways all other nontransformational grammars) misses is the fact that many strings of different sets are related to each other, in class sequence and in word choice. The center string *NtVN* (*Crowds demanded rights*), and the conjunctional right adjuncts *and tVN* (*and demanded rights*), and the noun adjuncts on the right *wh- tVN* (*which demanded rights*) and on the left *N-Ving* (*rights-demanding*), etc., are all closely related. Of course, all these facts can be stated about the strings, but they do not come out from the defined or immediately discoverable properties of the strings. In transformational analysis they are obtained directly.

Finally, all grammars based on sentences and not on discourse miss the few grammatical dependencies among separate sentences of a discourse, such as between a pronoun and its antecedent in a different sentence.

3.6. Recurrent dependence process: Sentence by word neighbors

The strings of 3.5 with their contiguous combinability characterize precisely the set of sentence forms of the language. This can be tested empirically by checking a great many sentences, and by showing that any-

---

[27] Instead of the operation of zeroing in adjuncts, one can have families of similar adjuncts restricted in complicated ways to word occurrences in the host string. This would be a purer string description.

thing that contravenes 3.5 is not a sentence. However, a recurrent process utilizing the analysis in 3.5 can be used for an additional result: to discover the sentence boundaries within a discourse. Just as the process proposed in 3.2 not only discovers the boundaries of morphemes but also shows that a segmentation into morphemes exists, so the process proposed in 3.6 not only discovers sentence boundaries but also shows that a segmentation into sentences exists.

We consider the word-class sequences which constitute sentences, and we think in terms of distinguishing in them all positive transitional probabilities from those which are zero.

We can now state a recurrent stochastic dependence between successive word classes of each sentence form in respect to the string status of each of those words.[28] In this hereditary dependence of the nth word class, w_n, of a discourse on the initial sequence $w_1, \ldots, w_{n-1}$, sentence boundaries appear as recurrent events in the sequence of word classes. This is seen as follows: It follows from 3.5 that, if a discourse D is a sequence of word (or morpheme) classes and x, y, are strings (defined as in 3.5) included in D, then:

a. the first word class of D is:

    (1a) the first word class of a center string,

or (2a) the first word class of a left adjunct which is defined as able to enter before (1a), or before (2a);

---

b. if the nth word class of D is:

    the mth word class of a string x containing p word classes, $p > m$

then the $n + 1^{th}$ word class of D is:

    (3) the $m + 1^{th}$ word class of x

or (4) the first word class of a right adjunct which is defined as able to enter after the mth word class of x,

or (2b) the first word class of a left adjunct which is defined as able to enter before (3), or before (4), or before (2b);

---

[28] Cf. N. Sager, Procedure for left-to-right recognition of sentence structure, Transformations and Discourse Analysis Papers, 27, University of Pennsylvania, 1960. We take the selection of a particular word class w_n in the nth word position of the sentence as the outcome which depends on the selections $w_1, \ldots, w_{n-1}$ in the successive preceding word positions. As before, we disregard the probability weightings of each outcome, and note only which outcomes have a non-negligible probability.

c. If the nth word class of D is: the last word class of a left-adjunct string x, where x is defined as entering before the mth word class of a string y

then the $n + 1^{th}$ word class of D is:

      (5)   the mth word class of y,

or (6)   the first word class of a left adjunct defined as able to enter before the mth word class of y, and such as is permitted to occur after x,

or (4c) the first word class of a right adjunct defined as able to enter after x,

or (2c) the first word class of a left adjunct defined as able to enter before (6), or before (4c), or before (2c);

---

d. if the nth word class of D is: (7)   the last word class of a right-adjunct string x which had entered after the mth word class of a string y which contains p word classes, $p > m$,

or (7') the last word class of a right-adjunct string x which had entered after (7)

then the $n + 1^{th}$ word class of D is:

      (8)   the $m + 1^{th}$ word class of y,

or (6d) the first word class of a right adjunct defined as able to enter after the mth word class of y, and such as is permitted to occur after x,

or (4d) the first word class of a right adjunct defined as able to enter after x,

or (2d) the first word class of a left adjunct defined as able to enter before (8), or before (7d), or before (4d), or before (2d);

---

e. if the nth word class of D is: the last word class of a right-adjunct string x which had entered after a string y

then the $n + 1^{\text{th}}$ word class of D is:

(1e) the first word class of a center string,

or (6e) the first word class of a right adjunct defined as able to enter after y, and such as is permitted to occur after x,

or (4e) the first word class of a right adjunct defined as able to enter after x,

or (2e) the first word class of a left adjunct defined as able to enter before (6e), or before (4e), or before (2e),

or (2e′) the first word class of a left adjunct defined as able to enter before (1e), or before (2e′),

or (9e) null;

---

f. if the nth word class of D is: the last word class of a center string
then the $n + 1^{\text{th}}$ word class of D is:

(1f) the first word class of a center string,

or (4f) the first word class of a right adjunct defined as able to occur after a center string,

or (2f) the first word class of a left adjunct defined as able to occur before (4f) or before (2f),

or (2f′) the first word class of a left adjunct defined as able to occur before (1f) or before (2f′)

or (9f) null.

Thus, we are given two possibilities for the string standing of the first word class of a discourse; and given the string standing of the nth word class of a discourse, there are a few stated kinds of possibilities for the string standing of the $n + 1^{\text{th}}$ word class. The possibilities numbered (2) are recursively defined. This relation between the nth and $n + 1^{th}$ word classes of a sequence holds for all word sequences that are in sentences, in contradistinction to all other word sequences.

We have here an infinite process of a restricted kind. In cases 1e and 1f, 2e′ and 2f′, 9e and 9f, the nth word class of D is the end of a sentence form,[29] and the $n + 1^{th}$ word class of D, if it exists, begins a next putative sentence form. The transitions among successive word classes of D carry hereditary dependencies of string relations. But 1e, f and 2e′, f′ are identical with 1a and 2a, respectively. The dependency process is therefore begun afresh at all points in D which satisfy 1e, f or 2e′, f′. These points therefore have the effect of a non-periodic recurrent event for the process.

In this way, sentence boundaries are definable as recurrent events in a dependency process going through the word classes of a dicourse; and the existence of recurrent events in this process shows that a sentential segmentation of discourses exists. Given a sequence of phonemes, we can then tell whether the first word is a member of a first word class of some sentence form, and whether each transition to a successor word is a possible string transition. That is, we can tell whether the phoneme sequence is a sentence. In so doing we are also stating the string relation between every two neighboring words, hence the grammatical analysis of the sentence in terms of string theory.

3.7. *Cycling cancellation automaton for sentence well-formedness*

The characterization of sentences as against nonsentences on the basis of the string analysis of 3.5 can be given another form, yielding a way of determining each sentence boundary different from that of 3.6, if we consider the sequence not of words but of string relations in the sentence. For this purpose, each word is represented by the string relations into which it enters. We then have an alphabet of symbols indicating string relations, and each word of the language is represented by a sequence of these new symbols. On any sequence of these new symbols we can, by means of a simple cycling automaton, erase each contiguous pair consisting of a symbol and its appropriate inverse, in the manner of free-group cancellation. Then each sequence which can be completely erased represents a sentence.

There are three string relations (i.e., conditions of well-formedness) that hold between words of a sentence: they can be members of the same string;

---

[29] This holds also if the $n + 1^{th}$ word class of D does not satisfy any of the possibilities for cases e, f, since only 6e, 4e, 2e, and 4f, 2f can continue a sentence form beyond a point where it could have ended. In this case the $n + 1^{th}$ word class of D is part of no sentence form. Aside from this, if the mth word class does not satisfy the table above, then D is covered by sentence-form sections only up to the sentence end preceding m.

or one word is the head of an adjunct and the other the host word to which the adjunct is adjoined; or they can be the heads (or other corresponding words) of two related adjuncts adjoined to the same string. Furthermore, since sentences are constructed simply by the insertion of whole strings into interior or boundary points of other strings, it follows that the above relations are contiguous or can be reduced to such. For each string consists of contiguous word classes except insofar as it is interrupted by another string; and each adjunct string is contiguous in its first or last word to a distinguished word of its host string (or to the first or last word of another adjunct at the same point of the host).

This makes it possible to devise the cycling automaton. In effect, we check each most-nested string (which contains no other string within it), i.e., each unbroken elementary string in the sentence. If we find it well-formed (as to composition and location in host), we cancel it, leaving the string which had contained it as a most-nested, i.e., elementary, string. We repeat, until we check the center string.

In view of the contiguity, the presence of a well-formed (i.e., string-related) word class B on the right (or: left) of a class A can be sensed by adding to the symbol A a left-inverse $b\`$ (or: a right-inverse $'b$) of $B$. If the class $A$ is represented by $ab\`$, the sequence AB would be represented by $ab\`b$; and the $b\`b$ would cancel, indicating that the presence of B was well-formed. For example, consider the verb *leave* in the word class V_{na} (i.e., a verb having as object either NA, as in *leave him happy*, or NE, as in *leave him here*). It would have two representations: $va\`n\`$ and $ve\`n\`$. Then *leave him happy* could be represented by $va\`n\`.n.a$,[30] which would cancel to v, indicating that the object was well-formed.

Specifically: For any two word classes (or subsequences of word classes) X, Y, if (a) the sequence XY occurs as part of a string (i.e., Y is the next string member to the right of X), or (b) Y is the head of a string which can be inserted to the right of X, or (c) Y is the head of a string inserted to the right of the string headed by X—then, for this occurrence of X, Y in a sentence form, we set $X \to y\`$ (read: $X$ is represented by $y\`$, or: the representation of X includes $y\`$) and $Y \to y$, or alternatively $X \to x$, and $Y \to 'x$.[31] Here $y\`$ is the left inverse of y, $'x$ is the right inverse of x, and the sequences $y\`y$, $x'x$ (but not, for example, $xx\`$) will be cancelled by the device here proposed.[32]

---

[30] The dots separate the representations of successive words, for the reader's convenience, and play no role in the cancellation procedure.

[31] Correspondingly, if Y is the end of a string inserted to the left of X, etc., then $Y \to x\`$, $X \to x$; or $Y \to y$, $X \to 'y$.

[32] In this notation, it will be understood that $(xy)\` = y\`x\`$, $'(xy) = 'y'x$, and $(x\`)\` = x = '('x)$.

It should be stressed that what is essential here is the fact that the inverse symbols represent the string relations of the word in whose representation they are included. When a most-nested symbol pair is cancelled, we are eliminating not a segment of the sentence but a string requirement and its satisfaction.

It follows that if a word class Z occurs as a freely occurring string, i.e., without any further members of its own string and without restriction to particular points of insertion in other strings, then its contribution to the well-formedness of the sentence form is that of an identity (i.e., a cancellable sequence). The representation would be $Z \rightarrow z`z$. (While such classes are rare in languages, an approximation to this in English is the class of *morevoer, however, thus,* etc.)

If, in a given occurrence in a string, a word class has more than one of the string relations listed above, its representation will be the sum (sequence) of all the string relations which it has in that occurrence. E.g., if in a given string, N has two adjuncts, A on its left and E on its right (as in *young men here*), its representation there should be $`ane`$; $a.`ane`.e$ would cancel to n.

There are certain conditions which the representations must meet. No two different classes should have the same representation (unless the string relations of one are a subset of the string relations of the other), for then we could cancel sentence forms that have one of these classes instead of the other. No proper part of a whole string should cancel out by itself, for then if only that part (or its residue in the string) occurred instead of the whole string, it would cancel as though it were well-formed. If a proper part of a string is itself a string (and so should cancel), then it should be considered a distinct string; otherwise the extra material in the longer string has the properties of a string adjoined to the shorter string. E.g., in $NtVN$ (*He reads books*) and NtV (*He reads*) we have two distinct strings, with appropriate representations for their parts.

Most word classes occur in various positions of various strings. For each of these occurrences there would be a separate representation, and the occurrence would be well-formed in a particular sentence if any one of these representations cancelled with its neighbors. As is seen in the next paragraph, for example, N has not one but several representations. We do not know which, if any, of these is the appropriate one for a given occurrence of N in a particular sentence until we see which, if any, cancels one of the representations of the neighboring words in that sentence.

Generally, host words carry the inverses of their adjuncts, as in $`ane`$ above for N; heads (as markers) of strings carry the inverses of the whole string (so that that the members of such strings do not have to carry the inverses of their next string member): e.g., in *which he will take* we can represent *which* by $wv'f'n'$, *he* by n, *will* by f, and *take* by v, so that after cancellation all that will be left is w to indicate that a well-formed *which* string occurred here. Order of adjuncts can be expressed as follows: In respect to noun adjuncts, N, TN, AN, TAN all occur, but not ATN (e.g., *the star*, *green star*, *the green star*, but not *green the star*). We have to give these classes, therefore, the following representations (in addition to others):

$$
\begin{array}{ccc}
T & A & N \\
t & a & n \\
 & 'ta & 'tn \\
 & & 'an
\end{array}
$$

Then the above examples would cancel down to n in the following sequences of representations: n, $t.'tn$ (*the star*), $a.'an$ (*green star*), $t.'ta.'an$ (*the green star*); but no representations could cancel *green the star* (ATN).

Permuted elements require separate representation. Thus *saw* as a verb requiring N object is represented vn'; but since the object can be permuted (as in *the man he saw*) the verb also has the representation $v'n$.

Some consequences of language structure require special adjustment of the inverse representation:

Delays. Sequences of the form $x'x'xx$ will cancel only if scanned from the right. If the scanning is to be from the left, or if the language also yields sequences of the form $yy'y'y$ which can only be cancelled from the left, then it is necessary to insert a delay, $i'i$, to the right of every left inverse (and to the left of every right inverse) which can enter linguistically into such combinations. We obtain in such cases $x'i'ix'xx$ and $yy'yi'i'y$, which cancel from either direction. This occurs in English when a verb which requires a noun as object (and the representation of which is vn'), or certain string heads like *which* (in a representation ending in n') meet a compound noun (the representation of which is $n.'nn$). The representation vn', for example, is therefore corrected to $vn'i'i$, so that, e.g.

$$
\begin{array}{ccc}
take & book & shelves \\
vn'i'i & n & 'nn
\end{array}
$$

cancels from the left (as well as from the right) to v.

Conjugates. There are also certain rare linguistic situations (including intercalation) which yield a noncancellable sequence of the form $z'x'zx$. An x' which can enter linguistically into such an encapsulated situation has to be representable by its Z conjugate $zx'z'$, which enables the x' and z to permute and the sequence to cancel. If a string AB, represented by $a\,'a$, encircles X (i.e., X is embedded within AB), once or repeatedly, then the relation of encirclement requires each A to be represented by xax': then, e.g., *AAXBB* yields $xax\backprime.xax\backprime.x.\backprime a.\backprime a$, cancelling to x. (Dots are placed to separate the representations of each capital letter.) In some cases, a particular word X (such as markers discussed below) requires the presence of particular other words or classes Y at a distance from it, i.e., with intervening Z, after all cancellable material has been cancelled: YZX occurs, but not ZX without Y. The representation of X then has to include $\backprime z\backprime yz$ so as to reach over the Z and sense the presence of Y. Thus the representation for *than* will check for a preceding *-er*, *more*, *less*; the representations for *neither*, *nor* will check for a preceding *not* (or negative adverb like *hardly*) or a following *nor*: *More people came than I had called*; *He can hardly walk, nor can he talk*.

Exocentrics. Another problem is that of a string XY which occurs in the position of a class Z, i.e., which occurs where a Z was linguistically expected: here the X would be represented by $zy\backprime$, so that $zy\backprime.y$ would cancel to z. (This XY includes grammatical idioms.)

Markers. If, in any linguistic situation requiring a special representation, one of the words occurs only in that situation or in few others, then that word can be treated as a marker for that situation, and all the special representations required to make the sequence regular can be assigned to the marker. This applies to the heads (initial words) of many right adjuncts, e.g., *that*, *whether* (*that he came*, etc.). This makes uncouth representations for the markers (e.g., for the *wh-* words), but it saves us from having alternative representations for more widely occurring word categories. Similarly, certain classes of verbs (verbs of ϕ_s in 4.2.1) occur as (adverbial) interruptions in sentences; e.g., *Celsius, I think, is Centigrade.* Rather than put the inverse of these verbs as an alternative representation of every noun (so that the noun plus verb interruption should cancel), we put the inverse of the noun onto the representation of these particular verbs: *think* is not only V *that*$\backprime$ (to cancel in *I think that …*) but also $\backprime nd$ (so that *I think* is $n.\backprime nd$ which cancels into an adverb, as proposed under Exocentrics).

Among the linguistic phenomena, in addition to the above, which were inconvenient for the inverse representation were:

Dependent repetition. The coordinate conjunctions permit recurrence of the preceding word class or sequence but not of most others (*N and N*, *V and V*, but not *N and V*). One or another of the participating words has to carry inverses that restrict the class (or class sequence) after *and* to being the same as the one before *and*; and we must allow for sequences which may intervene between the class and *and*. For example, we may give each class X a representation $xx\backprime + \backprime$.[33]

----

[33] In XY *and* XY, it suffices to provide for X *and* X, since each X cancels its Y. $+$ represents *and*.

Variable for permuted repetition. If a single symbol or string head X is repeatable, provision is made by allowing one of the representations of X to be xx or $'xx$. If there is a set Z of single symbols or string heads which may repeat in any order, as with certain right adjuncts of N, then various difficulties are avoided if we define a variable Z which takes as values zero and the inverse symbols for these right adjuncts, and which is placed between N and the inverse of each right adjunct. E.g., if an N has a following adjunct J, the N should be represented by nZj',[34] and after this had cancelled a following j, the Z could repeatedly take the value j', or w', etc., for any next following right adjunct of N. The maximum number of representations that this Z can indicate is necessarily less than the number of words following the given N in the given sentence. Thus one representation of *man* is nZw' for, e.g., *man who came* ... with any number of further right adjuncts on *man* after the *who came*: the Z is replaced by as many additional w', j' and other adjunct inverses as there are adjuncts after *man*.

Excision. Certain strings lose one of their words X in certain environments. Since the X is expected in that string, i.e., the string representation contains x', we have to add the x to the excision marker if there is one, or at some other appropriate point, in order to cancel the x'. If the marker is at the wrong end of the string for cancellation, e.g., to the left of x, we have to set up in addition to x' an alternative $'x$ which the marker on the left can cancel. Such is the case with the *wh-* markers. Thus one of the representations for *that, whom* is $wnv'f'n'$, so that we get:

the	*man*	*whom*	*he*	*may*	*see*
t	$'tnw'$	$wnv'f'n'$	n	f	$v'n$

which cancels to n for *man*, the rest being two adjuncts of *man*. To cancel, we had to use the $v'n$ representation for *see*, which allows for the object of *see* appearing before the verb (here the object is included in the *whom*, *that*).

One of the features of language which is not convenient for the inverse representations is the case of words which occur in a great many strings. Another is zero elements, whose string relations have to be included in the representation of neighboring word categories.

We can now state how the device operates: Given a sentence form, we replace each successive word class in it by the set of inverse representations of that class as determined above. We then obtain a set of representations of the sentence form by taking the cross-product of the class representations

---

[34] j represents the verb with suffix *ing*. nZj' would represent *people* in *people seeking help*, or *people seeking help who came to us*, etc.

for each successive class.[35] We now scan each sentence form representation in one direction (left to right or right to left), cancelling every sequence of the form $x\backslash x$ or $x\,'x$. Upon reaching the end of the sentence representation, we repeat this last process (i.e., we scan the reduced sentence representation, and cancel) until there is a scan in which no cancellation takes place. In this case no cancellation could occur in any further scans. The maximum number of scans for a sentence representation of n symbols is $n/2$. If now everything in the sentence representation has been cancelled, then the given representation (indicating a particular set of string relations of the classes in the sentence form) was well-formed; i.e., the sentence was well-formed for the represented string analysis (and hence meaning) of it. If, however, there is a nonempty residue in the sentence representation, then this was not the case.

In considering the amount of storage needed to hold all the representations of a sentence form, we note that some of the most frequent classes have only one representation, with two or so additional ones to allow for conjunction and adverbs (which could be represented otherwise if desired): f (tense; also $ff\backslash + \backslash$, $fd\backslash$, $'df$), t (article; also $'dt$), v_0 (verb having no object; also $v_0 v\backslash + \backslash$, $v_0 d'$). Every sentence (except imperatives and certain colloquial forms) has at least one f. Every word class in the sentence form replaces one word of the sentence, except that certain members of f are only suffixes.

The arbitrarily chosen complicated sentence in 3.7 cancels in seven scans. We have selected for each word class the representation which will fit into the cancellation, something that the device would have come by only in the course of scanning each sentence representation separately. The representations were not made especially for this sentence, but were selected from the list given in the full publication. The complicated representation for *one*, at the end, is a representation which each N has; it contains the representation of *wh-* words, and becomes useful when a *wh-* word has been zeroed (here, zeroed from *the one which they criticized*).

---

[35] Using only the set of representations in the table above, the sequence TAN would have the following representations: $t.a.n$, $t.'ta.n$, $t.a.'tn$, $t.'ta.'tn$, $t.a.'an$, $t.'ta.'an$; only the last of these would cancel to n. We do not try to have the device select those representations of each class which would match (cancel) the other classes in the given sentence form, for this would require the whole apparatus of string analysis. But that is not needed, since all string relations have already been expressed in the representation; so that all that is now needed is a local check of each representation. Properties of such representations as those of 3.5 and 3.7 are studied in A. K. Joshi and H. M. Yamada, String-adjunct grammars, Transformations and discourse analysis papers 75, University of Pennsylvania 1968.

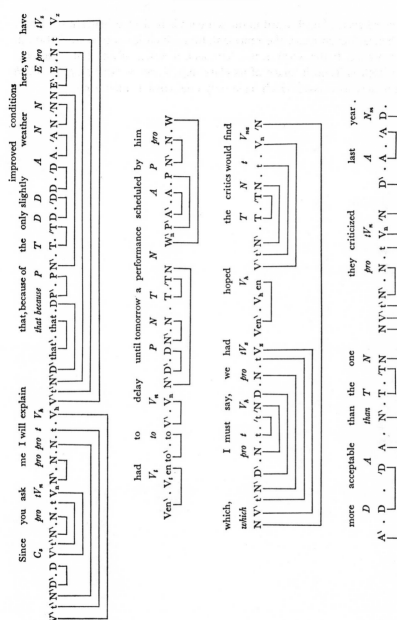

Figure 3.2

The word class of each word in the sentence is indicated beneath the word, and beneath each class, the representation which leads to cancellation in this sentence. If the word occurs here as a member of an idiom, then the word itself is listed in place of its class: e.g., *because*. (But *that, to, which* appear here as classes which have only one word as member.)

4

Sentence transformations

4.0. From sentence forms to sentence relations

From the assumption of the finiteness of the grammar, it follows that there must be a system of finite rules, some of which would have to be recursive, that suffice not only to characterize sentence forms (or discourse forms), as in Chapter 3, but also to characterize the actual sentences. Such a system of rules must specify which word values occur for the word-class variables in all the sentence forms. It must also be able to provide for the different degrees and kinds of sentence acceptability noted in 2.3. It is clear that the word choices, and the acceptabilities, can only be given for a finite subset of sentences; on this must depend the rules for all other sentences. The method which proves adequate for this purpose does not simply take all sentence forms and add restrictions to determine their word choices. Rather, it finds a connection between word choice and acceptability, on the basis of which it establishes a certain equivalence relation in the set of propositions (i.e., sentences equipped with an acceptability grading; 4.1.5.1). This equivalence relation enables us to isolate a finite set of elementary sentence forms whose choices of word values we characterize by an acceptability grading. From these graded elementary sentences we can derive the acceptability grading of all other sentences (or sound sequences). This completes the solution to the central problem of grammar. At the same time, as will be seen below, the method opens the way to many results about language, and about its mathematical properties.

The method starts with a theory of intersentence relations. This is an approach to grammar which asks primarily not how sentences are segmented (as do the methods of Chapter 3) but how sentences are related to other sentences. The basic relation that is established here is that which holds between sentence forms which are satisfied by the same word choices, in yielding acceptable sentences: for it turns out that the word choices which produce acceptable sentences are not unique to particular sentence forms, but common to several; and whereas the list of word choices is indefinite and readily changeable, the list of sentence forms which are satisfied by the same word choice is as definite and stable as in the rest of the grammar. It is shown that the word choices of all sentences are combinations of a finite set of word choices (the ones appearing in elementary sentences, from which all sentences are composed), and that the strong structural and semantic similarities mentioned above hold between sentence forms whose acceptable word choices are identical. As a byproduct of the intersentence relations we obtain a decomposition of sentences, as Chapter 3 provided decompositions of sentence forms.

One might think that transformational analysis imports into linguistics some unverifiable data (about word choices and acceptability) which structural linguistics had found avoidable or unnecessary. However, the data are inescapable if the actual sentences of a language are to be characterized; structural linguistics leaves these data to the last, and then leaves them untreated, because when this problem is left to the end it becomes unmanageable, if for no other reason than that the set of sentence forms (for each of which the word-choice data would have to be supplied) is not finite but denumerably infinite. By showing among sentence forms a stable relation in respect to word-choice acceptability, transformational analysis is able to restrict the word-choice problem to the finite set of elementary sentences, whence the word-choice acceptabilities are recursively extended by transformations to the denumerable set of sentences. In this way the problem is faced within linguistics, as it should be, but it is now of manageable proportions. The result is that transformational theory characterizes the actual sentences of the language instead of the sentence forms alone; or rather, given the set of elementary sentences, it characterizes the set of all recursively obtainable sentences.

We therefore consider now a description of language structure which we will call transformational analysis, because it starts with partial transformations from one subset of sentences to another subset having the same word choices. It is mathematically the most interesting, since the only elements in it are relations among subsets of sentences; and it is also the most powerful grammatically (i.e., it makes the greatest number of interesting distinctions), and the most sensitive semantically. Our first objective will be to characterize each sentence, in a nontrivial way, as the image of particular sentences or sets of sentences under particular mappings; and, if possible, to factor each sentence entirely into sentences, ultimately into prime sentences. To achieve this, we have to find some definable property which each sentence has to one degree or another, in terms of which, for each given sentence A, we could distinguish particular sentences as being the inverse image, i.e., the source or components, of A. What we find is not a direct property of this kind, but a system of inequalities which can characterize a particular sentence and which is preserved in its source or component sentences.

4.1. An equivalence relation among sentences

4.1.1. A set of inequalities

To introduce this system of inequalities, we note first (as in 2.3) that we cannot say for every word sequence that it is either definitely a sentence of the language or definitely not one. Very many word sequences are definitely sentences, e.g., *I am here*, and others are definitely not, e.g., *ate go the*. But there are also many word sequences which are recognized as metaphoric, or as said only for the nonce, i.e., as extended by the present speaker on the basis of "real" sentences, e.g., *Pine trees paint well*, from such sentences as *Spy stories sell well*. Other word sequences are recognized as containing some grammatical play, or some borrowing from another language: *He is yes the oldest in the room*. There are nonsense sentences which are on the verge of ungrammatical (because of word subclass): *Indisputability mailed relatedness*; and pure nonsense sentences (because of word choice): *The book chewed the cup*.

It is also easy to find a word sequence for which speakers of the language disagree, or readily alter their judgment, or even cannot decide, as to whether it is a grammatical sentence of the language or not: e.g., *He is very prepared to go*; *He had a crawl over the cliff*. There can be various degrees of marginality for various word choices in a form: e.g., the gradation in *He gave a jump, He gave a step, He gave a crawl, He gave a walk, He gave an escape*.

We see that in the set of all finite word sequences, the subset consisting of those which constitute sentences of the language is by no means well-defined. In addition to the ones which are definitely in or out of the set of sentences, there are a great many marginal cases, word sequences which are in different degrees and respects partially but not entirely accepted as sentences.

Instead of speaking of different acceptabilities for word sequences as sentences, we can speak of different sentential neighborhoods, for each one of the marginal sentences can be an acceptable sentence (separately, or under an operator which operates on sentences) in a suitable discourse; whereas the completely nonsentential word sequences, i.e., those which do not belong to sentence forms, appear without acceptability ordering, and only as subjects or objects of special predicates (5.4). For the grammatically marginal, the required neighboring sentences would be metalinguistic: *If one may use the word 'yes' correspondingly to the use of 'not', I would say He is yes the oldest in the room*. For extensions of the grammar, the neighboring sequences might give the source of the extension: *Just as you had a walk along the valley, he had a crawl over the cliff.* For nonsense, the neighborhood might be nonsensical in the same way, as in a fairy tale or dream: *All the dishes ran over to the desk-top, and suddenly the book chewed the cup.* Aside from the obvious cases of marginal sentencehood, many sentences are really to be found only in particular types of discourse, i.e., in the neighborhood of particular other sentences; and many of these would indeed be but dubiously acceptable sentences outside of such discourses. Thus *The values approach infinity* is normal in mathematical discourse but rather nonsensical outside it. *It rotates about the bond*, acceptable in chemistry, may be nonsense in ordinary English, where *bond* is an abstract noun from *bind*, or an item of finance.

The property of acceptability does not depend on the truth of the sentence: *The book is here* is a grammatically acceptable sentence even if the book in fact is not here. It does not even depend on meaning in any direct way. While there is a certain connection between meaningfulness, in the colloquial sense of the term, and acceptability, the connection is simply that a sentence which is acceptable in a given neighborhood has meaning in that neighborhood. We cannot assert that it is acceptable because it has meaning (i.e., because the combination of its word meanings makes sense) for in many cases a sentence is acceptable not on the basis of the meanings that its words have elsewhere. If *Meteorites flew down all around us* is an acceptable sentence, it is not because this is a meaningful word combination, since *flew* means here not "flew" in the usual sense but some movement distantly similar to it. The datum is simply that *flew* is used here, i.e., occurs in an acceptable sentence with *meteorite*, by virtue of some extension or other process of language use. Nor can we say that acceptability exists for every word combination which makes sense in any way, for there are word combinations which can make sense but are not acceptable sentences: *Man sleep*; *Took man book. The air swam* has very low acceptability, although it can have meaning in the colloquial sense: the air moved in or on the water with the motion used by fish, people, etc. But metaphors like *His ideas flew thick and fast* can be fully acceptable even though the meaning of *flew* outside such metaphors would not support a meaningful combination of that word with *ideas*.

The standing of a word sequence (with the required intonation) in respect to membership in the set of sentences is thus expressible not by two values, yes and no, but by a spectrum of values expressing the degree and qualification of acceptability as sentence, or alternatively by the type of discourse or neighborhood in which the word sequence would have normal acceptance as a sentence.

The complicated and in part unstable data about acceptability can be made available for grammatical treatment by means of the following pair test for acceptability: Starting for convenience with very short sentence forms, say ABC, we choose a particular word choice for all the word classes, say $B_p C_q$, except one, in this case A; for every pair of members A_i, A_j of that word class we ask how the sentence formed with one of the members, i.e., (1) $A_i B_p C_q$, compares as to acceptability with the sentence formed with the other member, i.e., (2) $A_j B_p C_q$. If we can obtain comparative judgments of difference of acceptability, we have for each pair A_i, A_j either (1) > (2) or (1) < (2) or (1) = (2) approximately.[1] In this case the relation is transitive, so that if $A_i B_p C_q > A_j B_p C_q$ and $A_j B_p C_q > A_k B_p C_q$ it will always be the case that $A_i B_p C_q > A_k B_p C_q$. We would then have a linear ordering of the members of A in respect to acceptability in $AB_p C_q$. If the judgments which we can obtain in this pair test are not quantitative but rather in terms of grammatical subsets (metaphor, joke, etc.) or language subsets (fairy tales, mathematics, etc.), then some reasonable ordering of the subsets could be devised so as to express the results in terms of inequalities (A_i being acceptable in the same subset as A_j, or in a subset which has been placed higher or lower). The results of the test might also take the form of some combination of the above. It is possible that we would obtain a relation in the set A such that for each pair either equality or inequality holds in respect to $AB_p C_q$, but without the relation being transitive. The system of inequalities, obtained from such a pair test, among sentences of the set $AB_p C_q$ will be called here a grading on the set $AB_p C_q$; each sentence $A_i B_p C_q$ which is graded in respect to the other sentences of the set will be called a proposition (4.1.5.1). There may be more than one grading over the A in $AB_p C_q$ (footnote 1).

Other gradings can be obtained for all members of A in respect to every other word choice in BC. In the same sentence form ABC we can also obtain a grading over all members of B for each word choice in A and C, and so for all C for each word choice in AB. We may therefore speak of a grading over the n-tuples of word choices in an n-class sentence form as the collection of all inequalities over the members of each class in respect to each word choice in the remaining $n - 1$ classes.

---

[1] We may obtain more than one answer as to the acceptability difference between $A_i B_p C_q$ and $A_j B_p C_q$. E.g., *The clock measured two hours* may be judged more acceptable than *The surveyor measured two hours*, in the sense of measuring off, i.e., ticking off, a period of time, but less acceptable than the latter in the sense of measuring something over a period of two hours. In all such cases it will be found that there are two or more gradings (in the sense defined immediately below) over all members of A in $AB_p C_q$; each grading is associated with a sharply different meaning of $AB_p C_q$ and, as will be seen later, a different transformational analysis of $AB_p C_q$ (see end of 4.1.2).

4.1.2. Inequalities preserved in sentence forms

It now turns out that, given the graded n-tuples of words for a particular sentence form, we can find other sentence forms of the same word classes in which the same n-tuples of words produce the same grading of sentences. E.g., in the passive sentence form $N_2 t$ be Ven by N_1 (where the subscripts indicate that the N are permuted, with respect to $N_1 tVN_2$), whatever grading was found for *A dog bit a cat, A dog chewed a bone, A book chewed a cup* is found also for *A cat was bitten by a dog, A bone was chewed by a dog, A cup was chewed by a book*. And if *The document satisfied the consul* may be found in ordinary writing, but *The partially ordered set satisfies the minimal condition* in mathematical writing, this difference in neighborhood would hold also for *The consul was satisfied by the document* and *The minimal condition is satisfied by the partially ordered set*.

In contrast, note that this grading is not preserved for these same n-tuples in $N_2 tVN_1$: *A cat bit a dog* is normal no less than *A dog bit a cat*; *A cup chewed a book* is nonsensical no less than *A book chewed a cup*; but *A bone chewed a dog* is not normal, whereas *A dog chewed a bone* is.

This need not mean that an n-tuple of words yields the same degree of acceptance in each of these sentence forms. The sentences of one of the forms may have less acceptance than the corresponding ones of the other, for all n-tuples of values or for a subset of them. E.g., *Infinity is approached by this sequence* is less acceptable than *This sequence approaches infinity*;[2] but if the second is in mathematical rather than colloquial discourse, so would be the first.

To summarize the empirical results:

1. Using any reasonable sense of acceptability as sentence and any test for determining this acceptability, we will find, in any sentence form $A(X_1, \ldots, X_n)$ of the variables (word classes) $X_1, \ldots, X_n$, some (and indeed many) n-tuples of values for which the sentences of A do not have the same acceptability, i.e., inequalities hold among them as to acceptability. Alternatively, given this A, we will find some n-tuples of values for which the sentences of A appear in stated different subject matter types of discourse.

2. Given this result we can then find (apparently always) at least one other sentence form $B(X_1, \ldots, X_n)$ of the same variables, in which the same n-tuples yield the same inequalities, for the same tested speakers; or

---

[2] And so for all $N_1 tVN_2$ whose form under the ϕ_s operators defined below contains $VnPN_2$ with P $\neq$ *of*, as in *the approach to infinity;* compare *He purchased books, Books were purchased by him, He made a purchase of books*, where P = *of* in the form under ϕ_s, and where the acceptability is not markedly lower in the passive.

alternatively in which the same *n*-tuples show the same difference as to subject matter types of discourse.

The precise degree of acceptance, and sometimes even the precise kind of acceptance or neighborhood, is not a stable datum. Different speakers of the language, or different checks through discourse, may for some sentences give different evaluations; and the evaluations may change within relatively short periods of time. Even the relative order of certain *n*-tuples in the grading may vary in different investigations.[3] However, if a set of *n*-tuples yields sentences of the same acceptability grading in *A* and in *B*, then it is generally found that the instabilities and uncertainties of acceptance which these *n*-tuples have in *A* they have also in *B*. The equality of *A* and *B* in respect to the grading of these *n*-tuples of values is definite and stable.

In this way, the problem of unstable word choices is replaced by the stable fact that the same (perhaps unstable) word choices which order the acceptability of sentences in one sentence form *A* do so also in some other one *B*.

A sentence may be a member of two different gradings, and may indeed have a different acceptability in each (e.g., *The clock measured two hours*, below). It will then have a distinctly different meaning in each. Some sentences, called ambiguous, have two or more common different meanings, which means that they have normal acceptance as members of two or more different graded subsets. Consider the ambiguous *Frost reads well* (in the form *NtVD*). We have one grading of a particular subset of sentences, one in which *Frost reads well* and *Six-year-olds read well* are normal, while *This novel reads well* is nonsense; this grading is preserved in the form *N tried to VD*: *Frost tried to read well*, *Six-year-olds tried to read well*, *The book tried to read well*. In another grading, of another subset of sentences, *Frost reads well* shares normalcy with *This novel reads well*, while *Six-year-olds read well* is dubious; this grading is preserved in the form *One can VND*: *One can read Frost well* (or: *smoothly*), *Reading Frost goes well*, *One can read this novel well*, *One can read six-year-olds well*.

The relation between two sentence forms in respect to preservation of acceptability grading of *n*-tuples is not itself graded. We do not find cases in which various pairs of forms show various degrees of preservation of acceptability grading. The recursive preservation of the gradings (4.0) is

---

[3] There are special cases involving language change or style and dialect difference. E.g., people may accept equally *They laughed at him*, *Students talk about the war*, but may grade *He was laughed at by them* more acceptable than *The war was talked about by the students*.

a consequence of this. In most cases, over a set of n-tuples of values, two sentence forms either preserve the acceptability grading almost perfectly or else completely do not. The cases where grading is only partially preserved (as in footnote 3) can be reasonably described in terms of language change, or in terms of different relations between the two forms for two different subsets of n-tuples.

4.1.3. Over specified domains

When we attempt to establish the preservation of the grading between two sentence forms, we find in some cases that it does not hold for all the n-tuples of values in both forms, but only for a subset of them. For example, the $N_1 t V N_2 / N_2 t$ be Ven by N_1 relation holds for *The clock measured two hours/Two hours were measured by the clock* but not for *The man ran two hours* ($\nexists$ *Two hours were run by the man*). We must then define the relation between the two forms over a particular domain of values of the variables. This raises immediately the danger of trivializing the relation: There are many more word n-tuples than sentence forms, hence for any two sentence forms of the same word classes it is likely that we can find two or more n-tuples of words whose relative grading is the same in the two forms. E.g., given only *A dog bit a cat, A hat bit a coat*, and *A cat bit a dog, A coat bit a hat*, we could say that the grading preservation holds between $N_1 t V N_2$ and $N_2 t V N_1$ for this domain. Interest therefore attaches only to the cases where a domain over which two forms have the same grading is otherwise syntactically recognizable in the grammar, e.g., because the domain covers all of a given morphological class of words (e.g., all adjectives, or all adjectives which can add -*ly*); or because the domain, or the set of n-tuples excluded from the domain, appears with the same grading also in other sentence forms and is too large a domain for this property to be a result of chance.

One can specify the domain of one grading preservation in terms of the domain of another. For example, the passive is not found for triples which appear with the same relative grading both in $N_1 t V N_2$ and in $N_1 t V$ *for* N_2 (for all N_2 of a given V):

> *The man ran two hours.*
> *The man ran for two hours.*

But: $\quad\nexists$ *Two hours were run by the man.*
> *The clock measured two hours.*
$\quad\nexists$ *The clock measured for two hours.*
> *Two hours were measured by the clock.*

Or one can specify the domain in terms of stated subsets of words. Some of the sets of n-tuples whose grading is preserved as above have some particular word, or an unextendable subset of words, or the complement set of these, in one of the n positions. For example, the passive is never formed from n-tuples whose V is *be, become*: $\not\exists$ *An author was become by him.*[4]

In other cases, a subset of words in one of the n positions is excluded from the domain of the grading preservation, but only in its normal use; the same words may occur in that position in a different use. In the preceding example, we can say that the passive does not occur if N_2 names a unit of duration or measurement (*mile, hour*) unless V is one of certain verbs (*measure, spend, describe*, etc.). But in many, not all, of these cases, it is not satisfactory to list a subset of words which is excluded from the domain of a certain grading preservation between two forms, because these words may be nevertheless found in the domain, but with very low acceptance. For example, *The clock measured for two hours*, excluded in the preceding example, may indeed occur, in the sense of *The clock measured something, over a period of two hours*. And for this sense, *The clock measured two hours* has indeed no passive: So *The clock measured two hours* has two distinct meanings, one (barely acceptable) which appears also in *The clock measured for two hours* (also barely acceptable), and one normal one which appears also in the normal *Two hours were measured by the clock*.

Finally, since the use of words can (with some difficulty) be extended by analogy or by definition, it is in many cases impossible to exclude a word from the domain of a grading preservation (what we will later call a transformation). Thus if someone were to use *run* in the sense of *run up* (as to run up a bill, or to run up a certain number of hours out of a time allotment), he might say *They ran (up) two hours* and this could have a passive *Two hours were run (up) by them*. But the passive would be only for this sense of *They ran two hours*, and not for the one which is a grading preservation (transform) of *They ran for two hours*. In general, then, the domain of a grading preservation is a set of n-tuples defined not by a list of words in a particular position, but by the fact that that set of n-tuples participates in stated other transformations.[5]

---

[4] It will be seen later (4.2.4) that most of the problems of restricted domains apply to an aberrant case: the analogic transformations. These domains are partly due to the composition of these analogic transformations as products of the base transformations. For the base transformations (4.2.1), the domains are unrestricted, or restricted in simple ways.

[5] Cf. Henry Hiż, Congrammaticality, batteries of transformations, and grammatical categories, *op. cit.* in Chapter 1. The two occurrences of *two hours* may be distinguished as having different classifiers, and the different domains may thus, in a regularized language (6.5), be distinguished as having different classifiers.

There are certain cases of unextendable word subclasses, e.g., the verb *be*, which can be extensionally listed. But this unextendability is itself a syntactic property.

The sets of word n-tuples which preserve grading in two sentence forms are therefore characterized in most cases by syntactic properties, as above. At the worst, the justification for distinguishing such a set of n-tuples is residually syntactic (e.g., to complete a set of grading-preserving form pairs). Each set of n-tuples, or rather the word class (in one of the n positions) which characterizes the n-tuple set will almost always be found to have a semantic property (e.g., the *be* verbs, the unit names); but semantic properties can often be found also for other word sets which do not play a role here.

We therefore require that the grading preservation hold not necessarily over all word n-tuples of a sentence form, but over a nontrivial syntactically characterizable subset of these. This reduction of our requirement is especially useful in the case where we can partition the n-tuples of a sentence form A into subsets $a_1, a_2, \ldots, a_n$ such that in each subset the grading is preserved as between A and another sentence form B; but the grading in the set of all n-tuples of A is not preserved in B because the order of the a_i subsets of A may not be fully preserved in B. Thus, if we considered all n-tuples in NtV we would not find their acceptability order preserved in $NtV_v aVn$ (where V_v is a set of aspectual verbs such as *try*, *give*, and Vn is a nominalization of the original verb, often by zero, as in *to glance/a glance*): The order is indeed preserved as between many word choices, e.g., *The man glanced*, *The doll glanced*; *The house glanced*; *The truth glanced* and *The man gave* (or: *threw*) *a glance*; *The doll gave* (*threw*) *a glance*; *The house gave* (*threw*) *a glance*; *The truth gave* (*threw*) *a glance*. Similarly for *The man smiled*; *The doll smiled*; etc., and *The man gave a smile*; *The doll gave a smile*; etc. And so on. But whereas *The man gave* (or: *threw*) *a party* is normal, *The man partied* is virtually unacceptable or very marginal. However, we note that within whatever low acceptability they have, *The man partied* (*all night*) is nevertheless more (or differently) acceptable than *The statue partied* (*all night*), which is more so than *The truth partied*; and this order is preserved in *The man gave a party*; *The statue gave a party*; *The truth gave a party*. Hence $NtV_v aV_i n$ preserves the ordering of NtV_i for each V_i within subsets of V.

The existence of restricted domains makes a general definition of transformations difficult; and there may be difficulties in discovering the precise domain of some particular transformations. However, the fact that many transformations have no restriction as to word subsets gives an initial stock of well-established transformations for a language which helps in determining various remaining transformations.

4.1.4. *Between a sentence and a sentence pair*

Hitherto we have considered sentences the grading of whose word-choices is preserved in some other set of sentences. However, for all sentence-forms A, except the relatively short ones, the grading–preserving relation will be found, not only as between A and some other sentence-form B, but also as between A and certain sets (pairs, triples, etc.) of sentence-forms C, D, ..., F, where the logical sum of the word-classes in C, D, ..., F equals those of A. Thus the form $N_3 N_2 t$ *be Ven by* N_1 (e.g., *Wall posters were read by soldiers*) has this relation not only to $N_1 t V N_3 N_2$ (*Soldiers read wall-posters*) but also to the pair $N_2 t$ *be Ven by* N_1, $N_2 t$ *be P* N_3 (*Posters were read by soldiers, Posters were on a wall*), or the pair $N_1 t V N_2$, $N_2 t$ *be P* N_3 (*Soldiers read posters, Posters were on a wall*). Similarly, the sentence-form N's *Vn* $V_{ss} N$'s *An* (as in *His arrival caused her lateness*) has this relation not only to N's *An* t *be* $P_{ss} N$'s *Vn* (*Her lateness was because of his arrival*) but also to the pair NtV, $N t$ *be A* (*He arrived, She was late*).

What makes this manageable is that it will be found that there is a small number of elementary sentence forms, and that a small number of (transformational) physical differences suffices to connect every sentence form to one or more of these elementary ones. All sentence portions which appear as substructures in pre-transformational linguistics (e.g. the compound noun *wall posters*) are relatable to sentences (e.g., *Posters are on a wall*) in a regular way and with preservation of grading for the word-choices involved.

4.1.5. *Transformation defined*

4.1.5.1. *As equivalence relation in the set of propositions*

We now take two sentence forms A and B, of a given n word classes or morpheme classes or subclasses (as variables), where A and B differ by some fixed morphemes or small sets of morphemes (as constants), or in the order or omission of certain classes, and where, in each form, the sentences produced by each set of n values for the variables of the form are graded as to acceptability or discourse neighborhood. Since a particular sentence form may have two or more gradings for the n-tuples of values of its variables, we will use the term propositional form for a sentence form equipped with a particular grading of the n-tuples. We define a transformation $A \leftrightarrow B$ between the two propositional forms A, B, over all n-tuples of their values or a syntactically characterizable domain of them, if the grading of the n-tuples is identical for the two. If we define a set of propositional sentences (propositions), where each proposition is a sentence with a position in a grading, and is identified by an n-tuple of values in a particular propositional form, then each transformation takes each proposition A_i of one propositional form A into the corresponding proposition B_i of another form B. Each grading preservation gives thus a partial transformation in the set of propositions.[6]

A modification of this definition can be made for the case of 4.1.4, where the relation holds between A and a set B of sentence forms.

Transformational analysis is thus not primarily an indicator of the structure of each sentence separately, but rather a pairing of sets $\{A\}$, $\{B\}$, of sentences, and so of the corresponding sentences A_i, B_i in each set, preserving sentencehood (approximate acceptability as sentence).

Transformations generate an equivalence relation in the set of propositional forms (and in the set of propositions), and impose a partition on them. Each proposition obtained by a given n-tuple in one form may be called a transform of the corresponding proposition obtained by the same n-tuple in the other form. If proposition A_1 is a transform of B_1, it occupies in the grading of $\{A\}$ the same position as B_1 in the grading of $\{B\}$. Transformations are thus defined not directly on sentences, but on propositions, i.e., graded sentences (sentences as members of a grading), or else, equivalently, on sets of sentences.

4.1.5.2. *As operation on the word sequences*

Given a transformation $A_1 \leftrightarrow A_2$, we now consider the difference between the morpheme sequences in A_1 and A_2. Since the various propositional forms which are transforms of each other contain the same word classes, the forms, as word class sequences, cannot be arbitrarily different from each other. For two transformationally related forms, the difference in sequence of words (or morphemes) and word classes is, in general: a permutation of word classes or constants; the addition or omission of a constant; and only in limited ways to be discussed below the addition or omission of a class. And since the individual sentences which are transforms of each other contain the same n-tuple of word values, they

---

[6] The term "propositional sentence" or "proposition" as defined here differs from "proposition" in logic, where it represents the set of all sentences that are paraphrases of each other. However, it is an approach in natural language to the "proposition" of logic, for paraphrases in language can be defined only on the basis of propositional sentences. Propositions will later be defined as particular sentence pairs (end of 5.1).

add no further difference to the above differences between the propositional forms.[7]

A transformational relation between two sets of propositions, A_1, A_2, can be described as a one-to-one mapping ϕ of A_1 onto A_2, in which each proposition in A_2 is identical with its inverse image in A_1 except for certain additions, permutations, or omissions which are associated with the mapping ϕ.

It is a fact of linguistics that given a transformation $A_1 \leftrightarrow A_2$ the permutations, additions, and omissions which differentiate a morpheme sequence (a proposition) in A_2 from its inverse image in A_1 are the same for all the propositions in A_2. This constant difference between propositions in A_1 and their image in A_2, which is associated with each transformation, will be called the trace ϕ_{21} of the transformation, and may be looked upon as an operation on the morpheme sequences of A_1 yielding those of A_2. We write

$$A_2 = \phi_{21} A_1$$

or:
$$\phi_{21} : A_1 \rightarrow A_2$$

or:
$$\phi_{21}{}^{-1} : A_2 \rightarrow A_1.$$

The importance of the trace is that it is a physical deposit in one member of the transformationally related pair of propositions. It will be seen that each proposition can be covered disjointly by segments, each of which is a transformational trace or a residual elementary sentence (which itself can be considered a transformational trace, 4.2.5). For this purpose, however, we have to accept as segments certain zero segments (traces of ϕ_z which erases the phonemic content of certain morpheme occurrences) and certain change-indicators (traces of permutational ϕ_p and morphophonemic ϕ_m); see 4.2.1. In English, these peculiar (non-incremental) segments are

---

[7] The one difference between sentences which goes beyond the difference between their corresponding propositional forms obtains if a word of an n-tuple has one form in one sentence and another in its transform, i.e., if the shape of a word changes under transformation. This is covered by the morphophonemic definition of the word; if the change is regular over a whole grammatical subclass of words, it is included in the morphophonemic transformations ϕ_m (below). A similar situation is seen when the regularity of morphemic difference applies in certain cases to families of syntactically-equivalent morphemes rather than to individual morphemes. Thus in some cases *-ing* and zero are equivalent nominalizing suffixes on verbs (imposed by certain ϕ_s operators): e.g., *He felt distrust* or *trust*, or *a dislike*, but *a liking*. At the same time, *-ing* occurs for all these verbs as a different "verbal noun" suffix (imposed by other ϕ_s operators): *He kept up his distrusting* or *trusting* or *disliking* or *liking*.

all due only to paraphrastic transformations, and not to the information-containing transformations. The same ϕ which are used as symbols for the transformational mapping between propositions can be used as symbols for the transformational segments which the mapping introduces in the image propositions. We have thus moved from transformations as a relation among propositions to transformations as a segmental decomposition of propositions.

For convenience in investigating the set of sentences under these transformational operators, we may define the positive direction of the operation in a consistent way for all $A_i \leftrightarrow A_j$ pairs. Here, two criteria are used, which apply equivalently for some pairs of related propositional forms, and complement each other in the remaining pairs:

1. We take the arrow in the direction $A_1 \rightarrow A_2$ if the number of sentence forms which include the sentence form or trace present in A_2 is smaller than the number of sentence forms which include that present in A_1: for example, we write *I say this* → *This I say* rather than *This I say* → *I say this* because the sentence nominalizations *Sn* do not exist for the *This I say* form ($\exists$ *My saying this*, $\nexists$ *This my saying*).

If as among these three forms the source A_1 were taken as *This I say*, then we would have

$$\exists \ \phi_{p1}: \quad \textit{This I say.} \rightarrow \textit{I say this.}$$
$$\exists \ \phi_{p'1}: \quad \textit{This I say.} \rightarrow \textit{This say I.}$$
$$\nexists \ \phi_{n1}: \quad \textit{This I say.} \rightarrow \textit{This my saying} \dots$$
$$\exists \ \phi_{np}: \quad \textit{I say this.} \rightarrow \textit{My saying this} \dots$$
$$\nexists \ \phi_{np'}: \quad \textit{This say I.} \rightarrow \textit{This saying my} \dots$$

If we accepted this source, we would have to say that nominalization ϕ_{n-} does not occur on the source but does occur on the p-permuted form and again not on the p'-permuted form. (ϕ_{n-} here is a symbol for the *Sn* portion of ϕ_s, 4.2.1.)

Whereas if the source A_1 is taken as *I say this*, we have

$$\exists \ \phi_{p1}: \quad \textit{I say this.} \rightarrow \textit{This I say.}$$
$$\exists \ \phi_{pp}: \quad \textit{This I say.} \rightarrow \textit{This say I.}$$
$$\exists \ \phi_{n1}: \quad \textit{I say this.} \rightarrow \textit{My saying this} \dots$$
$$\nexists \ \phi_{np}: \quad \textit{This I say.} \rightarrow \textit{This my saying} \dots$$
$$\nexists \ \phi_{npp}: \quad \textit{This say I.} \rightarrow \textit{This saying my} \dots$$

Here nominalization occurs on the source but does not form products with the permutations. This is clearly simpler than when *This I say* is taken as source.

2. We take the arrow in the direction $A_1 \rightarrow A_2$ if the number of morphemes (including reconstructible morphemes in the manner of 4.2.2.6) in the trace in A_2 is not less than in A_1.

In many cases the number of morphemes is observably greater: *It is old.* → *It is very old*; *It is old.* → *I think that it is old.* These are incremental transformations. In other cases the number of morphemes is the same, and the transformation consists in some other change: *He will come only now.* → *Only now will he come.* Finally, there are transformations in which the number of morphemes seems to decrease; however the formulation of zeroing (and pronouning) in 4.2.2.6 shows that the apparently lost morphemes are still present in the derived sentence, but in zero phonemic shape.

There are also other criteria which agree with these. E.g., we take the direction as $A_1 \rightarrow A_2$ if the discourse neighborhoods of A_1 are much more varied than those of A_2.

Transformations have (1) important uses in linguistics, and (2) a structure of their own: (1) They are important in linguistics because they show a relation among sentences which is not directly obtainable in other theories. Such relations are the partial structural similarities between certain sentences or segments of sentences (e.g., between *The man who attributed this to me was wrong.* and *The man attributed this to me.*), and also certain strong semantic similarities between sentences (e.g., between *The Air Force bombed villages.* and *Villages were bombed by the Air Force.*). The relations do not simply depend on similarity of word composition or grammatical structure,[8] and have far-reaching semantic interpretation. Transformations become more important when it turns out that in addition to being a relation which preserves sentencehood, transformations can indicate the structure of each sentence and that each sentence can be characterized by its transformational relations to a unique set of other sentences; and that indeed every sentence can be decomposed by transformations entirely into sentences. (2) Finally, their value appears even greater when it turns out that the transformations are not simply some *ad hoc* list of operations on sentences, but products of a few quite reasonable base operations. Since the properties (1) can be treated more precisely when the transformations are defined in respect to their structure (2), we first present the structure of the set of transformations (4.2) and then their properties (4.3,4).

---

[8] Thus *He painted two hours* and *He painted for two hours* are transforms of each other, but *He painted two boys* and *He painted for two boys* are not. (*Hours* here is in a different subclass of N than is *boys*.) *The artist's painting was of two boys* and *Two boys the artist painted* are transforms of *The artist painted two boys;* but *Two boys painted the artist* and *The boy painted two artists* are not.

4.2. *Elementary differences among sentences*

It was seen in 4.0 that some recursive rules for word choices and accept-ability, as well as for sentence forms, are unavoidable in language. There is nothing in principle which limits the complexity and number of these rules (so long as they are finite), or their relation to each other. It is an empirical fact, however, that the transformations which are found in a language satisfy very stringent conditions. It is possible to set up a small number of families of base operators in such a way that all the base operators within a family are physically similar to each other (differing in most cases only by choice of word within a word subclass), and that almost all transformations of the language can be obtained as successive applica-tions (products) of these base operators; there remain some unresolved transformations, which one may hope will eventually be shown to consist of some product of base operators. The analysis into base transformations has much explanatory value, because many seemingly unique facts in a grammar are found to be due to the juxtaposition of properties of the component base operators. These base operators turn out to be a reason-able set, i.e., they are what one might expect to need in making a language; and each makes a clear and reasonable semantic contribution to the ultimate sentence. Furthermore, the families of base operators seem to be much the same in all languages, as far as has been seen to date, so that we have here a universal item of language structure.

4.2.1. *Base operators*

We now consider how, starting with an equivalence relation on sets of graded sentences, we arrive at a system of base operators, or elementary differences among sentences. Consider a few transforms and the differences between them:

A_1:   *He speaks English*

A_2:   *He is a speaker of English*

A_3:   *He is speaking English*

A_4:   *He begins to speak English*

A_5:   *He is beginning to speak English*

ϕ_{21}:   *... be a __ -er of ...*

ϕ_{31}:   *... be __ -ing ...*

ϕ_{41}:   *... begin to ...*

ϕ_{51}:   *... be beginning to ...*

ϕ_{32}: $(\ldots \text{__} \textit{-ing} \ldots)(\ldots a \text{__} \textit{-er of} \ldots)^{-1}$

and so on.

We note that certain constants appear in many traces. We may suppose that if a constant, e.g., *-ing* appears in two or more traces, e.g., ϕ_{31} and ϕ_{51}, then there exists some trace which consists of *-ing*, and both ϕ_{31} and ϕ_{51} are products that contain this trace as a component. Furthermore some traces can clearly be obtained by the successive application of two or more other traces. In ϕ_{32} we obtain A_3 from A_2 by undoing part (or all) of ϕ_{21}, then adding part (or all) of ϕ_{31}. ϕ_{51} can be obtained by applying ϕ_{41} to A_1 (obtaining A_4) and then applying ϕ_{31} to the resultant A_4: $\phi_{51}A_1 = A_5 = \phi_{31}A_4 = \phi_{31}\phi_{41}A_1$. We therefore try to show that many traces are composed of other traces, and to find if possible a base set of elementary traces. If this situation were not the case, a language would have a great number of independent traces, each present in an independent transformational pairing of sentence sets.

Considerable analysis of the empirically found traces is necessary in order to show that certain traces are composed of others. This requires even more reformulation of the sentence differences and their domains than does the initial specifying of the transformations. The most powerful tool for this purpose, as will be seen below, is the defining of a zeroing operator (4.2.2.6) which reduces to zero certain words in stated neighborhoods, if certain general syntactic conditions are satisfied. For example, in *I insist on my getting first place* the zeroing operator yields *I insist on getting first place*; but in *I insist on his getting first place* the zeroing cannot operate. The dropping of *for* before nouns of duration (4.1.3) was another example of such zeroing.

The perhaps surprising result is not only that the traces in a given language can all be obtained as products of certain minimal traces, but also that these minimal traces are of very few types in each language, and rather similar as between one language and another. Each minimal trace can be considered as due to a base operator, which acts on sentences of all or of particular forms. Each operand form consists of particular ordered word classes or subclasses; each trace consists of additional such material concatenated with the operand, or else of changes in relative position or phonemic shape of morphemes in the operand. For English, it is sufficient to recognize the following general types of base operators:

ϕ_a: *word → expanded word*, e.g., *large → very large*. These are the base adjuncts. In every language there seem to be some words y which can be adjoined (on the right or on the left) to members of a particular word class x, in such a way that xy or yx appears in the same grammatical conditions as does x. We may say that xy or yx is a base expansion of x, and y a modifier of x. These modifiers are of a general (more or less quantitative) semantic character; and it turns out that the acceptability inequalities of sentences containing xy, yx are the same as for the corresponding sentences with x: compare the acceptabilities of *The box is large*, *The air is large* with those of *The box is very large*, *The air is very large*. Such y may therefore be considered transformations on the sentences containing x. There are many nonbase expansions, i.e., modifiers which can be transformationally derived (4.2.3) from operators on whole sentences (ϕ_s, below; see 4.2.2.1) or on pairs of sentences (ϕ_c).[9] Only base expansions, which cannot be derived from ϕ_s or ϕ_c are included here in ϕ_a.

$\phi_v : NtV_i\,\Omega_i \rightarrow NtV_v\,V_i\,a/n\Omega_i$. These are verb operators, such as cannot be derived from ϕ_s: *He studies* → *He is studious*, *He is studying*, *He is a student*, *He has studied*.

$\phi_s: S \rightarrow NtV_{-s}\,Sn$; $SntV_{s-}\,\Omega$. These are sentence operators: *John fell* → *I wonder whether John fell*; *That John fell surprised me*; *John's falling surprised me*.

Every language has a variety of such operators. Since these operators have a sentence as operand, and adjoin some material to it, the differences among operators of this family can be: (a) as to the kind of "nominalizing" deformation Sn which they impose on the operand S (e.g., *that John fell*, *John's falling*, both from *John fell*); (b) as to what the operator itself consists of; (c) as to dependences between the subject or object of the operator verb (V_{-s}, V_{s-}) and the subject or object of the operand sentence (e.g., *I prefer that he should go*, and *I prefer to go* ← *I prefer that I should go*; but for the operator *try* the subject must be the same as that of the operand S: *I try to go* ← presumed *I try that I should go*.). In each language, the varieties of these operators are a selection out of the possible combinations of those syntactic conditions; and their meanings depend upon this syntactic selection. A brief survey of the main varieties for English is given in 4.2.2.3.

$\phi_c: S_1, S_2 \rightarrow S_1 CS_2$. Connectives: *John appeared*, *I arrived* → *John appeared after I arrived*; S_1 here is called primary, and S_2 secondary. These are operators on two sentences, and are found in every language. Here too there are a few varieties, which differ as to the kind of similarities that are required between the two sentences of the operand. The essential syntactic (and semantic) varieties are: two coordinate C: *and*, *or*; and one or more varieties of subordinate C which express substantive connections (e.g.,

---

[9] E.g., the fact that adverbs of manner (such as *slowly*) cannot be adjoined to *is* follows from the fact that they are derived from an operator which operates on a form of nominalization which is not available for *is* (4.2.2.1): *He speaks French slowly.*← *His speaking of French is slow.* is analyzed as *slow* operating on nominalized *He speaks French*. But *He is a lord.* cannot be nominalized into (∄) *His being of a lord*, hence ∄ *His being of a lord is slow*, ∄ *He is a lord slowly*.

because, after). Connections among three or more sentences are obtained from various combinations of these binary connectives.

ϕ_p: Permutations (a few stated types) on the symbols above: *He will speak only here.* → *Only here will he speak.* These occur only within particular sentence forms. They have the effect of contrast with related sentences of the discourse; and also of simplifying the recognition (or computation) of the transformations. Thus, if the secondary CS_2 must contain a word present in S_1, that word is permuted to the beginning of S_2:

(*I successfully avoided him.*) *wh-* (*You mentioned him.*) →
(*I successfully avoided him.*) *wh-* (*Him you mentioned.*) =
I successfully avoided him whom you mentioned.

ϕ_z: Zeroing of determinable material (i.e., disappearance of its phonemic content) under ϕ_s and ϕ_c; only a few stated types. Zeroing is found in all languages and makes possible great and intricate reductions in the length of sentences. The main type of zeroing applies to repeated material in stated positions: *He came and he went* → *He came and went.* A second major type is the dropping of words which are unique to their operators or operands: *I found the book which you lost.* → *I found the book you lost.* The third, in English, is the dropping of disjunctions of nouns covering a whole class, such as could be represented by indefinite pronouns: *He opposes X's or Y's ... or Z's drinking* → *He opposes anyone's drinking.* → *He opposes drinking.* In all these cases, the zeroing is carried out only if specified syntactic conditions are satisfied, and the zeroed words are recoverable (up to synonymity) from the remaining sentence. It thus becomes possible to say that words and morphemes are never lost; their shape becomes zero. Under somewhat different conditions, pronouning occurs instead of zeroing. In 5.6 a generalization of zeroing will be proposed, which makes possible a considerable regularization of conjoined sentences.

ϕ_m: Morphophonemic change, of the phonemic shape of a morpheme under an operator, or next to particular other morphemes: present tense (*t*) on *be* yields *is*; past tense (*t*) on *be* yields *was*, etc.

Successive application of these seven families of base operators yields all the transformations of English, with the exception of certain ones which have not yet been resolved. In English, ϕ_v, ϕ_s, ϕ_c, ϕ_z suffice by themselves to produce the great bulk of transformational relations, and so of sentence forms. The seven operators were arrived at empirically, in a search for a base set for the numerous transformational relations. However, it is evident that they are a reasonable set of operators, and bring understandable syntactic and semantic changes into the sentences on which they operate.

The first four operator families ϕ_a, ϕ_v, ϕ_s, ϕ_c introduce particular meaning-bearing increments into their operand. The last three do not, and their resultant always paraphrases their operand. In the above list, the following word classes and sequences appeared, and are distinguished by their operators or their operands:

S: sentence. Elementary sentences have the form $NtV_i\,\Omega_i$.

N: simple (concrete) nouns, not obtained from Vn, Sn, e.g., *book*.

t: the tense morphemes for past and present, and with each of these possibly an auxiliary: *will, may*, etc.[10]

V: concrete verbs, collected into certain subclasses V_i according to the Ω_i which they require.

Ω_i: verb object consisting of i, where i ranges over zero, N, PN (preposition plus N, e.g., *in town*), A (adjective, e.g., *good*), D (adverb, e.g., *well*). Each V_i requires the corresponding Ω_i to follow it; e.g., *exist* (V_0) requires zero (*It exists*), *wear* (V_n) requires N (*He wears a hat*).

V_v: Certain verbs (of aspectual character) not generally included in V_i; grouped in various subclasses according to the traces, here marked a/n, which they impose on their following V: e.g., *be* (of *be -ing*), *have* (of *have -en*).[11] *Not* can be considered an aberrant subclass of V_v.[12]

Va/n: Verb which has been made adjectivelike, Va (e.g., *receptive, receiving, to receive*, from *receive*), or nounlike, Vn (e.g., *a walk* from *walk*), or very rarely remains unchanged (e.g., *do* in *I'll make do without it*, from *I'll do without it*). These changes in the verb are to adapt it for use as follower of V_v: *He is receptive to it, He began receiving it, He wants to receive it, He took a walk.*

V_{-s}, V_{s-}: Certain verbs or predicates (intellectual, emotional, causative, etc.) not generally included in V_i or V_v; grouped in various subclasses according to the traces, here marked n, which they impose on the following or preceding S, respectively.

---

[10] The auxiliaries can be analyzed as members of t, or else as operators on t; *not* is an operator on all these. These operators can be considered as aberrant subclasses of V_v—aberrant primarily in that *not* does not precede t (as does the first V or V_v). For more detail see footnote 12.

[11] If a word has different local synonyms in the different word classes listed here, it will be considered to be a different word in each of its synonym sets. Thus *will* in *will go* (synonym *shall*) is not the same word as *will* in *wills to go* (where it is a concrete V, synonym *desire*). Thus *have-en* (of *have taken*) and *take* (of *take a walk*) are different words than the *have* (of *have a book*) and *take* (of *take a book*) which are in V_n.

[12] Aberrant in that *not* requires no change in the following V; and even more so in that *not* accepts no suffixes: hence past and present tenses do not move to become suffixes of *not*, and a V_v preceding *not* places the a/n suffix on the V following *not* instead of on *not* (e.g., *be-ing* on *leave* yields *be leaving*; *be-ing* on *not* on *leave* yields *be not leaving*; while *be-ing* on *try to* on *leave* yields *be trying to leave*).

Examples of V_{-s}:

> *know* (*I know that he left, I know of his leaving*, etc.),
> *hope* (*I hope that he left*),
> *wonder* (*I wonder if he left, about his leaving*),
> *cause* (*This caused him to leave, his leaving*);

of V_{s-}:

> *surprise* (*That he bought books surprised me*),
> *is a fact* (*That he bought books is a fact, His buying books is
>   a fact, His buying of books is a fact*),
> *is frequent* (*His buying books is frequent, His buying of books
>   is frequent*),
> *is slow* (*His buying of books is slow*).

Sn: The resultant of certain increments to S, required by ϕ_s on that S, as seen in the parentheses immediately above. Sn is called a nominalization of S. E.g., from *The man arrived*:

> *that the man arrived,*
> *whether the man arrived,*
> *the man's arriving,*
> *the arrival of the man.*

The various increments enable Sn to satisfy the requirements of various subclasses of V_{-s} for a follower (i.e., particular kinds of nominalized sentence as object) and the requirements of various subclasses of V_{s-} for a preceder (i.e., particular kinds of nominalized sentence as subject):

> *I know that the man arrived.*
> *That the man arrived is a fact.*

C: Two major types of connectives: coordinate C_o: *and, or*; subordinate C_s: *since, because*, etc. To these may be added comparative C_p: *less than*, etc.; relative C_w: *which*, etc. All these differ from each other in the types of S on which they act and in the similarities which they require between the two S, and also in the permutations and zeroings which can be carried out on the resultant SCS. The C_w and C_p can be shown to be transforms of particular C_o, C_s with special equating and comparing sentences, so that only two types of connective are independent: C_o, C_s.

4.2.2. *Details of the base operators*

In order to show how the great wealth of transformations in a language can be obtained as successive applications of operators from the seven base types above, we note here the chief types of operators for English in each base family, and we sketch the justification for reducing the main transformations of English to these. The mass of detail in 4.2.2 has no other interest here, and may be omitted if the reader is satisfied to assume that the base operators described above indeed account for the transformations of the language. Some of the derivations seem complex and unexpectedly circuitous; but this does not alter the fact that for each transformation of the language (with the exception of a few unresolved cases) we can find a succession of the base operators whose final trace is identical with that of the transformation. The complete detail and special cases in English are of course far more than shown here, but entirely within the following framework.

4.2.2.1. ϕ_a: word expansions

In English, the great bulk of words which we might think of as being operators on (modifiers of) words turns out to be derived from operators ϕ_s on sentences (4.2.2.3). There are, however, a few transformations which cannot be stated in the form of ϕ_s, namely as operators on a sentence. These must therefore be considered as a separate set of operators on words. Among these are certain adverbs (of verb or of adjective):

> *He is strict, He is very strict.* $\nexists$ *His strictness is very.*
> *I forgot, I quite forgot.* $\nexists$ *My forgetting was quite.*

In the case of *I forgot, I simply forgot*, the form *My forgetting was simple*, is not transformationally related to *I simply forgot*.

In contrast, the great bulk of adverbs can be derived from ϕ_s:

> *He read the poem slowly.* ← *His reading of the poem was slow.*
> *He read poetry frequently.* ← *His reading poetry was frequent.*
> *He certainly forgot.*   ← *That he forgot is certain.*

One might think that all adverbs should be taken as originally operating on the verb, with any form like *His reading poetry was frequent* being derived therefrom. However many properties of adverbs follow automatically if the source is taken to be the ϕ_s form, but not otherwise. For example the adverbs of manner (e.g., *clearly*, *slowly*, but not *frequently*) do not occur on *be*, or on scale verbs (*costs*, *weighs*, etc.), or on other adverbs of manner. If we look at the ϕ_s form, we find that adverbs of manner impose on their operand sentence a particular form of nominalization *Sn*, namely one in which an *N* object has *of* before it:

> *His reading of poetry was slow.*
> ∄   *His reading poetry was slow.*
> ∄   *That he read poetry was slow.*

We now note that the sentences to which an adverb of manner cannot be added are precisely those which cannot be nominalized in this way. Thus *He is a fool* cannot be nominalized to *His being of a fool*, and indeed ∄ *His being of a fool is slow* (or any other adverb of manner) and ∄ *He is a fool slowly*. Similarly, from *The book costs $5.*, *The panda weighs 20 pounds*, we cannot obtain the nominalization with *of*: ∄ *The book's costing of $5.*, ∄ *The panda's weighing of 20 pounds*, and so ∄ *The book's costing of $5 is slow*, ∄ *The book slowly cost $5* (and so for any other adverb of manner).[13] In contrast, adverbs of occurrence (e.g., time) require in their operand sentence only the weaker nominalization without *of* (though the form with *of* is secondarily possible):

> *Our reading poetry was frequent in those days.*
> *His writing such letters is a recent activity.*

This nominalization is available for *be*, *costs*, etc., hence we have

> *His being a fool is frequent*, and hence *He frequently is a fool.*
> *The panda's weighing 20 pounds was recent*, and hence *The panda recently weighed 20 pounds.*

Somewhat similarly, other words which seem to be operators on words can be derived as morphophonemic forms (ϕ_m) of operators on sentences. E.g., for *He reads only English* there seems to be no sentence-operator form: ∄ *His reading of English is only*. However it can be shown that *He reads only English* is a transform of *He reads English and he doesn't read non-English*, so that *only* is obtained by ϕ_m operating on ϕ_c (conjunctions) with particular distribution of *not* in the second sentence.

And most adjectives adjoined to nouns will be also derived from ϕ_c: *He found a small book.* ← (*He found a book.*) *wh-* (*The book is small.*). If the adjective is taken as a direct addition (ϕ_a) to the sentence, it would not meet the grading–preservation conditions for a transformation, since the effect of the adjective on the acceptability of a sentence differs for different word choices: e.g., *He blew some small puffs* is as normal as *He blew some*

---

[13] The fact that this nominalization with *of* does not exist for these verbs is itself due to elementary properties: (a) *The book costs $5.* is derivable from *The book is $5. in cost.*, and so on; (b) in the structures on which the *of*-nominalization is defined, *be* is not an original member but has been added by a morphophonemic operator.

puffs, but in *He blew some small air* is unacceptable while *He blew some air* is normal. However, if the adjective is taken as derived from ϕ_c, grading is preserved: *He blew some small air* is as acceptable as the lesser of *He blew some air* and *The air is small*. (See 5.6.2 for measuring by the lesser acceptability.) Many apparent ϕ_a can therefore be obtained, by ϕ_z and ϕ_m from ϕ_s and ϕ_c.

Nevertheless, there remain, apparently in each language, some ϕ_a, i.e., some operators on words which cannot be derived from operators on sentences, or at least cannot be so derived as long as we stay within the set of actual sentences and do not use the more powerful regularization of Chapter 6. As for all ϕ, the operand of ϕ_a is a sentence, not merely a word, for it is the occurrence of a word in a sentence. E.g., in *Some boy saw a boy*, ϕ_a: *some* operates not simply on *boy* but on *boy* in a particular sentence position.

4.2.2.2. ϕ_v: *operators on verb*

As with ϕ_a, so also most transformations whose trace is of the ϕ_v form are derivable from ϕ_s by zeroing (mostly of the subject) in the operand *Sn*: *He prefers working* ← *He prefers his working* which is a case of ϕ_s (*he prefers*) on *S* (*He works.*). However, there remain certain cases of the ϕ_v trace in which no zeroing could have occurred:

has V en:	*He has gone, He has slept.*
is V en:	*He is gone, He is drunk.* (on few *V*).
is V{a}	*He is receptive to ideas.* (on few *V*).
is V er:	*He is an observer of mores.*
is PV{n}P:	*He is in love with her.*
is N{a}:	*This is problematic.*

(Here {a} and {n} indicate sets of adjectivizing and nominalizing suffixes, respectively. *P* indicates preposition.)

All these are clearly transforms of the corresponding sentences without the ϕ_v trace: *He goes, He sleeps, He goes, He drinks, He receives ideas, He observes mores, He loves her, This is a problem*. To these sources, each of these ϕ_v adds an aspectual meaning concerning the relation of the subject to the action of the verb (termination, proneness, etc.). Whereas ϕ_s, as will be seen below, makes the whole operand *S* into what is called the subject or object of the operator in the resultant sentence, ϕ_v makes just the verb with its object from the operand *S* into the object of the operator in the resultant sentence. Zeroings under ϕ_s, however, can make the resultants of ϕ_s look like resultants of ϕ_v. Hence those ϕ_s whose subject must be the same as the subject of their operand (end of 4.2.2.3) can be taken as ϕ_v, and will be so taken in various derivations given below.

4.2.2.3. ϕ_s: operators on sentence

This is the most complex family of operators. Each subset of these operators imposes a particular deformation on the operand S, thus making the operand S into a nounlike subject or object of the operator (as in *I know that John takes it*, or *John's taking it is strange*), or into a form similar to a subsidiary clause (as in *I requested John to take it* where the operand S is *John takes it*). In English there are six types of such deformation, from S (i.e., $NtV\Omega$) to:

1. *that S*;
2. *whether S_i* [*or not S_i*] (square brackets here and below indicate optional omission);
3. *for N to VΩ*; *that N* [*should*] *VΩ*;
4. *N's Ving Ω*; *Ving Ω by N*;
5. *N's Ving of Ω*; *Ving of Ω by N* (here and in 6, *of* appears only if Ω begins with *N*);
6. *N's Vn of Ω*.

In the forms below, a number before S (and in subscripts, the number after $_s$) indicates the deformation of S as numbered above.

Each operator consists either of a verb with its object, of which the deformed S is subject, or else of a subject (in most cases, N_h, the humanlike subclass of N) and verb (possibly with some N as object) of which the deformed S is object or subsidiary clause.

A great many types of ϕ_s can be constructed from different combinations of these possibilities. The main ones for English are:[14]

$1S\,V_{s1-N_h}$:	*That he purchased books surprised me.*
$1S$ is N_{s1}:	*That he purchased books is a fact.*
$1S$ is A_{s1}:	*That he purchased books is true.*
$2S\,V_{s2-N_h}$:	*Whether he purchased books* [*or not*] *intrigues me.*
$2S$ is N_{s2}:	*Whether he purchased books* [*or not*] *is the problem.*

---

[14] S here is *He purchased books*. Only one member is given from each operator type. The subscripts indicate subclasses of the operator words, but some subscript details which are required for the given type are omitted for convenience. For a detailed analysis of these operators in French, cf. Maurice Gross, *Transformational analysis of French verbal constructions*, Centre national de la recherche scientifique, Paris (Transformations and Discourse Analysis Paper 74, University of Pennsylvania 1967).

$2S$ is A_{s2}:	*Whether he purchased books [or not] is uncertain.*
$3S$ is A_{s3}:	*For him to purchase books is easy.*
$4S V_{s4}$:	*His purchasing books has already occurred.*
$4S$ is A_{s4}:	*His purchasing books is frequent.*
$5S$ is A_{s5}:	*His purchasing of books is slow.*
$N_h V_{-s1} 1S$:	*I know that he purchased books.*
$N_h V_{-s1} N1S$:	*I tell you that he purchased books.*
$N_h V_{-s2} 2S$:	*I wonder whether he purchased books [or not].*
$N_h V_{-s2} N2S$:	*I ask you whether he purchased books [or not].*
$N_h V_{-s3} 3S$:	*I require for him to purchase books.*
$N_h V_{-s5} 5S$:	*I imitated his purchasing of books.*

The meaning of each type of ϕ_s is related to the deformation it imposes. Deformations 1, 2, and 3 can be transformed into 4, and 4 into 5, but not otherwise; hence *His purchasing books surprised me* and *His purchasing of books surprised me* and *His purchasing of books is frequent*, but ∄ *That he purchased books is frequent*, and ∄ *His purchasing books is slow*. The operators that take different deformations consist mostly of different words: e.g., few words are members of both V_{-s2} and V_{-s3}. However, within a given deformation requirement, an operator may transform as between V, N, A, by ϕ_v: *Whether S is the problem* → *Whether S is problematic.*

In certain operators there is a dependence on the operand: one of the N in the operator must be the same as one of the N in the operand. These operators have characteristic meanings of their own, related to the kind of dependence; and the second occurrence of the common N is in many cases zeroed. (See end of 4.2.2.2.) The main forms are:

$N_i V_{-s4i} 4S$(with N_i as subject):	*He began his purchasing books.* → *He began purchasing books.*
$N_{hi} V_{-s4hi} 4S$(with N_i subject):	(?)*He tried his purchasing books.* → *He tried purchasing books.*
$N_i V_{-s5i} 5S$(with N_i subject):	*He did his purchasing of books today.* → *He did some purchasing of books today.* (A quantifier is stylistically helpful here.)
$N_i V_{-s6i} 6S$(with N_i subject):	*He made his purchase of books.* → *He made a purchase of books.*
$N_{hi} V_{-s3i} N 3S$(with N_i subject):	*He promised me that he would purchase books.* → *He promised me to purchase books.*

$N_h V_{-s3j} N_j 3S$(with N_j subject):   (?)*I forced him that he should purchase books.* → *I forced him to purchase books.*

$N_i V_{-s6io} 6S$(with N_i object):   (?)*He suffered their defeat of him.* → *He suffered defeat [at their hands].*

4.2.2.4. ϕ_c: connectives

The main properties of the subclasses of the conjunctions C, aside from the unique permutations and zeroings on them specified below, are:

C_o is commutative and associative: $S_1 C_o S_2 \leftrightarrow S_2 C_o S_1$ (*he will go and I will go, I will go and he will go*); but many cases of *and* are derived from C_s, which does not have this commutativity (*He took sick and died*). In addition, *but* may be obtained from *and* with certain meta-sentences.

C_s is neither commutative nor associative. It can, by products of the base operations (and their inverses, 4.2.4), be transformed into a form like ϕ_s, and thence to adverbial and other forms. Some examples:

> *He left because they shouted,*
> *He left because of their shouting,*
> *His leaving was because they shouted,*
> *His leaving was because of their shouting,*
> *Their shouting caused his leaving.*

In the last sentence above, the C_s has been replaced by a verb V_{ss} which connects the two S. Thus not all connectives ϕ_c consist of conjunction words C.

C_p requires both S to be of the form N *is* A, or requires both S to be transformed by means of identical transformations into Σ *is* A forms,[15] such as the following:

> $NtV\Omega \rightarrow$ *The quantity* (etc.) *of N's Ving* Ω *is great*
> $NtV\Omega \rightarrow$ *The number* (etc.) *of N who tVΩ is great*;

e.g., *It is longer than it is wide, He is taller than she is, The number of people who came was greater than the number who refused to come.* All further transformations are made on both S identically, yielding for example, *More people came than refused.* C_p are like C_s in having a ϕ_s form, e.g., *His height is greater than* (or: *exceeds*) *hers*; C_p are like C_o in permutation and zeroing.

---

[15] Σ represents N or Sn (with possibly various zeroings) as subject of *is* A.

It is possible to derive the comparative sentences from a fixed-form elementary metasentence of comparison

$$N_{deg} V_{comp} N_{deg}$$

where V_{comp} is the set of two verbs *equal, exceed* (which have various properties, such as accepting adverbs of degree but not of manner), and N_{deg} is a double occurrence of a degree noun (*amount, number, degree*); the remaining components of the comparative sentence are attached to this by C_w. Thus (6.8):

It is longer than it is wide.

would be derived by C_n and ϕ_z, ϕ_m from:

1. *F exceeds G.*
2. *It is F* (*inches*, etc.) *long.*
3. *It is G* (*inches*, etc.) *wide.*

C_w These are the words *who, which, where*, etc. (replaceable in certain cases by *that* as pro-word,[16] and by zero). The connective is *wh-* and the *-o, -ich, -ere*, etc., are pro-words for *N* or *PN*. In $S_1 C_w S_2$, the S_2 always begins (at least after permutation) with N_i (which may be *Vn* or *Sn* or any nominalized word) or PN_i or NPN_i, where N_i is some *N* which occurs in S_1. The *-o, -ich, -ere*, etc., are pro-words of this N_i or PN_i.

wh: *I found the man, The man called you → I found the man who called you.*

wh: *I bought a book, The book just appeared → I bought a book which just appeared;*

wh: *We painted that room, In that room he stayed → We painted that room where he stayed;*

wh: *I have a book, A page of the book he tore out*[17] *→ I have a book, a page of which he tore out.*

It is possible to derive the *wh-* conjunctions: *wh-* with comma from *and*, and *wh-* without comma from certain C_s (e.g., *if*), with a fixed-form metasentence of sameness comparable to that of comparison above (5.7.2).

---

[16] x is a pro-word of the word class or subclass X if x is a member of the synonym set (6.4) of each member of X. Thus *that* is a pro-word of *who, whom, which, what* (as well as being a member of other classes). Similarly, *he* is pro-word of the masculine subclass of *N*.

[17] This is obtained by ϕ_p from *He tore out a page of the book*. For the presence of *the* in S_2, see 5.7, 4.2.2.6 end.

4.2.2.5. ϕ_p: permutations

In general, permutations occur in a sentence only after some trace has been added to it. The main ones in English are the following:

$NtV\Omega \rightarrow \Omega NtV$ (*I suspect this* → *This I suspect*);
rarely $N\Omega tV$ (*I this suspect*), ΩtVN (*This suspect I*).

This permutation may be considered to be really a ϕ_m (i.e., change of shape) of *I suspect this in particular*, or the like.

The reason for claiming a hidden prior trace (here: *in particular*) is as follows: When ϕ_p operates on some trace (i.e., not on an elementary sentence), it is paraphrastic, as in:

I little expected this. → *Little did I expect this.*

When ϕ_p operates on an elementary sentence, i.e., where there is no overt sign of a transformational trace, it has a contrastive meaning, as in (1) *This I suspect*, etc. This is peculiar, because all other nonincremental operators are paraphrastic. For each of these contrastive ϕ_p there exist in the language paraphrases which consist of the addition of an adverb or a *CS* to the elementary sentence, as in (2) *I suspect this in particular*. Using the method of 5.6, we reconstruct the longer form (2), which contains the trace of some increment, as source, and obtain the short form (1) by a permutation with attendant recoverable zeroing. Then this ϕ_p is seen to be paraphrastic, as are all other ϕ_p and also the other nonincremental transformations (ϕ_z and ϕ_m); and it is seen that ϕ_p acts only on sentences containing a trace (not on elementary sentences), as do the other nonincremental transformations.

$NtV\Omega C_s S \rightarrow NtVC_s S\Omega$; $NtC_s SV\Omega$; $NC_s StV\Omega$; $C_s SNtV\Omega$:

I will think about the problem while driving →
I will think, while driving, about the problem;
I will, while driving, think about the problem;
I, while driving, will think about the problem;
While driving I will think about the problem

$NtV\Omega C_s S \rightarrow C_s StNV\Omega$: *Only because he's out of money will he return home* (preferably with certain *D*, e.g., *only*, before the C_s).

In $S_1 C_w S_2$, the $C_w S_2$ is normally permuted to immediately after the distinguished N_i in S_1 (C_w is generally *wh-*, the N_i in S_2 is pronounced as the second half of the *wh-*word):

wh: *This man dropped in, This man we know* → *This man dropped in, whom we know*; *This man whom we know dropped in.*

wh: *This man brought a book, This man went unnoticed* → *This man who went unnoticed brought a book.*

When the secondary S_2 is N_i is A or N_i is PN, the is, is P is often zeroed with the wh-word; the remainder of S_2, if short, is almost always permuted from the right of N_i in S_1 over to the left of N_i: A plate which is blue fell. → (∄) A plate blue fell. → A blue plate fell. Given repeated wh, as in S_1 wh S_2 ... wh S_n, the permuting of remainders of the secondary S has a certain order, based on the acceptability of A and PN sequences, so that certain classes of secondary S remainders are permuted to the left nearer to the N_i in S_1 and others farther to the left: A small blue plate fell. (A blue small plate fell. is not normal except due to further operators.)

After zeroing occurs in $C_o S_2$, $C_p S_2$, the CS remainder may be permuted leftward, but not beyond the last word in S_1 whose corresponding word was not zeroed in S_2.[18] (The subscripts below indicate whether a word is part of S_1 or part of S_2; not only the omitted words but also the bracketed ones are zeroable.)

> I_1 can$_1$ make$_1$ it$_1$ and I_2 will$_2$ make$_2$ it$_2$ →
> I_1 can$_1$ make$_1$ it$_1$ and $[I_2]$ will$_2$;
> I_1 can$_1$ and $[I_2]$ will$_2$ make$_1$ it$_1$.
> The$_1$ archaeologists$_1$ would$_1$ attribute$_1$ more items$_1$ to$_1$ the Etruscans$_1$ than the$_2$ historians$_2$ would$_2$ attribute$_2$ to$_2$ the$_2$ Etruscans$_2$ → The$_1$ archaeologists$_1$ would$_1$ attribute$_1$ more items$_1$ than the$_2$ historians$_2$ would$_2$ $[attribute_2]$ to$_1$ the$_1$ Etruscans$_1$.

4.2.2.6. ϕ_z: zeroing, pro-wording

The zeroings are restricted in an even clearer way. In general, morphemes which can be determined on the basis of the remainder of the sentence, and which occupy the position of adjuncts (3.5) in the sentence (more precisely: which are not the NtV of a primary sentence) can have their phonemic composition reduced to zero.

1. Zeroing of constants. A word or a whole CS may be zeroed in a given position in a sentence if it is a constant of a transformation, i.e., a distinguished segment in all occurrences of a particular transformation. Such a zeroing is the dropping (4.2.2.5 end) of the wh-word and the following is, after N: The man who is here → The man here. (But nothing drops if the V is not a constant; The man who came here doesn't change.) As an extension of this, we may assume that the zeroed words were any local synonyms of the constant; we may call them the appropriate words

---

[18] Under C_p, it is permuted not to the left of the word in S_1 which receives the more, less, as (which is the first part of the C_p).

for the given neighborhood. Thus in compound nouns N_2–N_1, which are derivable from N_1 *which has to do with* N_2 or the like, we can replace the constant *has to do with* by a local synonym suitable to the ordered pair N_1, N_2: e.g., *The milk bottle broke* ← *The bottle which is used* (or: *made*) *for milk broke* (with permutation of the post-*wh* remainder as above); *The milkman is here* ← *The man who deals with* (or: *delivers*) *milk is here*; but *The child who spills milk cried* ↦ *The milk-child cried.*[19]

A special case of the zeroing of material which is constant to a transformation is the zeroing of operators whose presence is marked by certain ϕ_m (especially intonation) or ϕ_p which operate only upon these operators. Thus, given

I like books.

(by $\phi_z \phi_c$) → *I like books and not other things.*
(ϕ_m: contrastive stress) → *I like books' and not other things.*
(or: by ϕ_p) → *Books' and not other things I like.*

we can zero the particular ϕ_c: *and not other things* since the contrastive stress or the permutation indicates its morphemic presence even when its phonemes are zero (4.2.2.5).

An interesting case of this type is the zeroing of performative-like ϕ_s, namely ϕ_s which make their resultants true. The effect here is that the meaning after zeroing is the same as before zeroing. If a person says " I promise you that I shall go " (but not " I promised you " or " He promises you ") then the statement is itself the stated promise, and its meaning is the same as that of " I shall go." There are reasons within transformational theory for considering that *I shall V*Ω can be obtained by zeroing from *I promise you that I shall V*Ω. Somewhat similarly, if we consider such a sentence as

The bombing was brutal, not to say brutish.

we find that the only way to account for the *not to say* within the base operators is to assume as source

I say that the bombing was brutal [even] if I am
not [going] to say that it was brutish.

The zeroing of the second *I* (which is needed as subject of *say*) can only be accounted for on the assumption that a first *I* had preceded. This first *I*

---

[19] But if the spilling was so habitual and known to the participants in the discourse that one could say *The child who is the one who has to do with milk (by spilling it) cried*, we might zero the verb sequence for this discourse to obtain *The 'milk-child' cried.*

must have had a verb after it, and both must have been zeroable. The only way to account then for the zeroing of the first I and its following verb is to assume that they constituted a ϕ_s : *I say that* which was zeroable as a performative ϕ_s. Thus S can be obtained by zeroing from *I say that S*. If we want to avoid having both S and *I say that S* as a source for S, we can say that $S \leftarrow I$ *say that S* only when S contains some trace of the zeroing of an *I say that*.

As a final example, it can be shown (4.2.3) that all questions $S?$ (e.g., *Is he coming today? When is he coming?*) are derivable from *I ask you: S?* (in turn by ϕ_m from *I ask you whether S*). After S has received the question intonation when under the ϕ_s: *I ask you*, the ϕ_s becomes determinate from the intonation. Furthermore, the meaning of the intonation on the operand S is the same as the meaning of the ϕ_s, since it is unique to that ϕ_s. In these conditions, the performative ϕ_s: *I ask you* is zeroable, and can always be recovered on the basis of the remaining intonation.

In analyzing SCS we will find justification for an important extension of the zeroing of constants of a transformation, i.e., the zeroing of sentence segments which are uniquely appropriate to the transformation. This extension begins with an understanding of appropriate words as those words in a given position of a given discourse which are the obvious ones to occur in that neighborhood, and are hence determinable on the basis of the neighborhood. We then extend appropriate zeroing to apply to whole CS which state standard definitions of words, or other information which is obvious to speakers of the language or to the participants in the given discourse. Justification for this last claim, and examples, will be given in the discussion of SCS structures (5.6.2).

2. Zeroing by antecedent. A word in a sentence which is under ϕ_s or ϕ_c may be zeroed (or, in particular conditions, pronouned) if the same word occurs in a stated position of the ϕ_s operator or the primary sentence S_1; the position in question depends on the subset of ϕ_s or ϕ_c.

Under ϕ_s and many C_s: adjunctlike forms (*for N, N's*) of subject N, if they refer to the same individual as an N of the operator or of the S_1, are zeroed optionally, or necessarily (depending on the subclass of V_{-s}, C_s); chiefly:[20]

$$N_i t V_{-s}(N_j)[\text{for } N_i] \text{ to } V\Omega$$
$$N_i t V_{-s}(N_j)[N_i's]Ving \; \Omega \text{ (or: } Vn\Omega)$$
$$N_i t V_{-s} N_j[\text{for } N_j] \text{ to } V\Omega$$
$$N_i t V_{-s} N_j[N_j's]Ving \; \Omega \text{ (or: } Vn\Omega)$$

---

[20] In the sentence forms, parenthesized segments are optional (i.e., two forms are being here summarized in one formula); bracketed segments are zeroable. The semantic demand that the words refer to the same individual is made syntactic in 5.7.2.

e.g., *I want [for me] to be first, I prefer [my] being first, I promise him to go (that I would go), I offered him to come later* (i.e., *he or I would come*), *I offered him [his] escape.*

$$N_i \, t V (N_j) C_s [\, N_k] \, t V \Omega \qquad k = i, j; \; C_s = \text{-}ing$$

I invited him, standing there by the door (i.e., *he or I stood*).

The replacing of a noun by pronoun (rather than by zero) is required if a noun of ϕ_s is repeated (referring to the same individual) in the operand of ϕ_s, or if in $S_1 C S_2$ a noun in a later occurrence or in S_2 repeats one in the other sentence. For the $S_1 C S_2$ case we therefore have, from *John will come if John can*:

> *John will come if he can.*
> *If he can John will come.*
> *If John can he will come.*
> ∄ *He will come if John can.*

For the ϕ_s case, from *John thought that John is safe* when both are the same person, the nonpronouned form does not exist and from the pronouning conditions we have

> *John thought that he was safe.*
> ∄ *He thought that John was safe.*
> *He was safe, John thought.*
> ∄ *John was safe, he thought.*

but ∃ *John was safe, or so he thought,* for here the *John* which is pronouned to *he* is in S_2 of the ϕ_c: *or.*

The conditions above are not violated by

> *His friends think that John is safe.*

because this is composed of

> *(Friends think that John is safe) wh- (Friends are John's),*

where *John's,* pronouned to *his,* is in S_2 of the ϕ_c: *wh-.* Since under ϕ_c the pronouning can take place in the later occurrence of the noun even if it is in S_1, we also have:

> *John's friends think that he is safe.*

Further under C_s and C_p, $V\Omega$ can be zeroed in S_2 (or in the second occurrence of $V\Omega$) if they are the same words as the corresponding $V\Omega$: *I will buy a book if you will buy a book* → *I will buy a book if you will; If you will buy a book, I will. The boy who wanted to go couldn't.*

Under C_o and C_p, we zero (usually only optionally) words in S_2 which repeat words in the same position of the same (or stated equivalent) ϕ in S_1.[21] That the zeroed word and its antecedent, from which it is zeroed, must be of the same ϕ can be seen, e.g., in

He took a walk and a ride ← *He took a walk and he took a ride,*

where *took* in both S_1 and S_2 is ϕ_v acting on *He walked, He rode*; but in *He took an umbrella and he took a walk*, where *took* in S_1 is V of the elementary sentence (4.2.3), there is no zeroing of *took* in S_2 (where it is ϕ_v): ∄ *He took an umbrella and a walk.*[22]

3. Zeroing of disjunctions. In a long disjunction of S in which one position takes each (acceptable) word value of the word class or subclass in that position, while the rest of the positions are unchanged from sentence to sentence, the disjunction of all members of a class (collected into one position by the C zeroing above) is replaced by zero (or by indefinite pronouns). E.g., *I heard English spoken* ← *I heard English spoken by N_1 or N_2 ... or N_n* ← *I heard English spoken by N_1 or I heard English spoken by N_2 ... or I heard English spoken by N_n.* Here $N_1, N_2, \ldots, N_n$ are all the possible subjects of *speak English* (at least, all relative to the given discourse).

In these three explicit ways the zeroed words are determined and hence informationally recoverable: either they are trivially well known relative to a transformation (as constants) or to the discourse (as appropriate words), or they are repetitions of an antecedent, or they are disjunctions of all words that can occur in the given position. The fact that the word has been zeroed is recognized from the absence of a word in the given position of the structure, e.g., the absence of *who is* from *The man here left*, the absence of *I ... make it* from the second part of *I can make it and will*, the absence of the subject of *speak English*. There is a subsidiary necessary condition for zeroing, which is satisfied in almost all cases: the zeroings are carried out only if the resultant abbreviated sentence still has the word-class form of a sentence (i.e., if it has the sequence of word classes and adjoined items that appear in sentences, although no longer a sequence

---

[21] There are a few restrictions which are due to conditions of ϕ operation too detailed to be presented here, e.g., under C_p, Ω must be zeroed (*He is taller than she is. He bought more books than she bought.*); but under C_o, Ω in most cases is not zeroed unless permutation occurs (*He bought, but did not wear, a ceremonial hat.* But: *He bought a ceremonial hat, but did not wear it.*). Also: If *wh*-remainders have been permuted to the left of an N in S_2, only certain such remainders permit the N to be zeroed (*He had six prints and I three. The small box fell but the large didn't.* But ∄ *I have a small new radio and he has a large old.*)

[22] ϕ here refers also to the elementary sentence-making ϕ_k (4.2.3).

of word subclasses that occurs otherwise as a sentence). Thus *I heard English spoken* is of the form of *NtVN* plus *Va* (adjective made from verb) adjoined to the *N*, somewhat like *I saw people arriving*; *I can make it and will* has the form of a sentence (*I can make it*) with something conjoined to the sentence (although the internal structure of what is conjoined has been reduced from the original and *I will make it*); and in *I can and will make it* the conjoined item has been permuted as allowed above. In contrast, a word is not zeroed if the result would create a type of word-class sequence which does not occur otherwise as a sentence form: *This bottle is (used) for milk* does not become *This bottle milk*. This is one of the almost-everywhere rules (7.1.2.3), and has exceptions such as the question (above).

The mapping of the set of zeroable sentences onto the set of sentences containing zeroing is a homomorphism. In a given sentence with zero in the nth position, the words there may have been zeroed for different reasons; i.e., the zero-bearing sentence may have different sources. E.g., *I favor swimming here* may have been zeroed from *I favor my swimming here* or *I favor N_1's or N_2's ... or N_n's swimming here* (= *anyone's swimming here*). The result is then an ambiguous sentence, derived by ϕ_z from two or more propositions.

Aside from zeroing, words (or disjunctions or conjunctions of the members of a word subclass) can be replaced by pro-words: *N* by *one*; *the N* by definite pronouns *it*, *he*, *she*, and *-ich*, etc.; *A* in certain positions by *so*; disjunctions by indefinite pronouns (*someone*, *some*, etc.). The conditions for replacement by pro-word are similar but not identical, to the conditions for replacement by zero. Aside from the usual pronouns above, we could use the methods of Chapter 6 to extend the derivation of pro-words so as to account for quantifiers and the like which would otherwise have to be considered as primitive adjuncts produced by ϕ_a. These can be considered as pro-words of certain subsets of quantifying and demonstrative adjectives, introduced upon their being permuted to the left (e.g., after zeroing of *wh- is*). Examples of such extreme possibilities are:

> *the*: e.g., *the book* ← *a book which is identified*
> (or: *unique*, etc.)
> *very*: e.g., *It is very wide* ← *Its width is great*
> (or: *considerable*, etc.).

4.2.2.7. ϕ_m: morphophonemics

Morphophonemic changes ϕ_m have to be recognized in order to obtain all sentences of English from the sentences produced by the base operators above. These ϕ_m are changes which alter the physical (phonemic) form of certain words or morphemes when they are brought by the above operations (or by well-formedness) into certain new neighborhoods. These morpho-phonemic changes are for the most part automatic, i.e., they necessarily operate when the new neighborhood is entered, but some of them are optional. They can be eliminated from the record of sentence-building (or sentence-relating) operators by being included, as operator-determined variant phonemic forms, in the physical description of each word or morpheme (when it first appears as part of elementary sentences, or of an operator). If ϕ_m are not taken as transformations, we have to allow the resultants of ϕ upon sentences to be not always sentences, and in the latter case the morphophonemic changes would be a finite mapping from these resultants onto the set of sentences.

Zeroing, and less easily pro-wording, could be considered simply a special case of morphophonemics, i.e., simply a zero shape for the word or morpheme affected, were it not for the fact that certain further operations such as permutations and analogic operations (4.2.4) can take place only after zeroing has taken place. There are cases when such an operator acts differently on certain sentence segments when some of their words have been zeroed than it does otherwise: e.g., N *which is* A is best described as being first zeroed (to NA) and only then automatically permuted to AN. Hence we have products of zeroing and these operators, so that the zeroing has to be listed in the set of operators. However, automatic zeroing is included in ϕ_m: e.g., if S_1 *wh-* $S_2 \leftarrow S_1$ *and* S_2 *and* S_{meta} (5.7.2), the replacement of *and* ... *and* S_{meta} by *wh-* is ϕ_m. In any case, only the para-phrastic ϕ_p, ϕ_z, ϕ_m and analogic ϕ ever depend on an occurrence of ϕ_z; the incremental ϕ never does.

4.2.3. *Products of elementary operators*

4.2.3.1. *Resultants; ambiguity from degeneracies*

All transformational relations among English sentences can be expressed as products (i.e., successive applications) of elementary differences (traces) listed under ϕ_a, ϕ_v, ϕ_s, ϕ_c, ϕ_p, ϕ_z, ϕ_m. This is not say that they have been derived from the base operators during the history of the language, or that the formulation of a sentence by a speaker (or its recognition by a hearer) follows these paths (see Chapter 8).

The residual operand of these base operators in all English sentences is the set of word sequences which are classifiable in the form $NtV_i \, \Omega_i$, which we will call the elementary sentence form S_e. Each operator acts on a sentence of this form (or on a pair of them), or on the resultant of one of these operators, to produce a resultant sentence, usually of this form. Except for a few (in any case, a finite number of) aberrant idiomatic sentences (such as *Handsome is as handsome does*), and except for various special problems which remain at this writing, all English sentences (including questions, imperatives) are of the two- to six-word form $NtV_i\,\Omega_i$ or are obtained from this form by a product of the base operators in the seven types above, including the irregular products of 4.2.4.

A proposition S_n, i.e., particular grammatical meaning of a sentence S_n, which is a transform of some S_k, and in particular of some residual sentences S_e, can be characterized by the sequence of base operators which carry S_k, and in particular S_e, through various intermediate transforms, to S_n. Thus for *I returned upon meeting him*, we have

(S_e)	(1) *I returned, I met him.*
(by ϕ_c)	→ (2) *I returned upon my meeting him.*
(by ϕ_z)	→ (3) *I returned upon meeting him.*

A derivation, i.e., a proof of the transformational derivability, of a sentence S_n from a sentence S_k, $k < n$, (ultimately from $S_1 = S_e$) is a sequence of base operators, $\phi_{k+1}, \phi_{k+2}, \ldots, \phi_n$, such that the operand S_m of the $m + 1^{\text{th}}$ operator is the resultant of the mth operator, for each m, $k \leq m < n$, and $\phi_n S_{n-1} = S_n$; S_1 is the resultant of the elementary sentence-making ϕ_1.[23]

When we wish to resolve a transformation between two sentence sets $\{S_n\}$ and $\{S_k\}$ into a product of base operators which derives each sentence in $\{S_n\}$ from the corresponding sentence in $\{S_k\}$, we may use various methods and indications. If the difference between $\{S_n\}$ and $\{S_k\}$ contains the trace of some base operator A, we try to resolve $\{S_n\} \leftrightarrow \{S_k\}$ into a product containing A. Thus, given that *Upon his signing the letter ..., I know of his signing the letter, His signing of the letter was meticulous, He is signing the letter*, are all transforms of *He signs the letter*, we would like to find, if possible, a single base operator which is responsible for the trace *-ing* in all these transforms. This may, surprisingly, turn out to be impossible in the present case, especially in view of the fact that the verb domain of *is—ing* is only a proper part of the verb domain of the other traces: $\nexists$ *He is knowing English*, $\nexists$ *He is having signed the letter*, but $\exists$ *I know of his knowing English*, $\exists$ *I know of his having signed the letter*. Identity of restriction in domain is an even stronger reason for considering that two transformations have a common component in base operators.

---

[23] This is a different sense of sentence derivation from that defined by Noam Chomsky in his generative grammar; cf. *op. cit.* (Chapter 1).

This consideration connects *He is in process of [his] signing the letter* to *He is signing the letter*; note ∄ *He is in process of knowing English,* ∄ *He is in process of having signed the letter.* (This may make it possible to derive *is—ing* from the other *-ing*.) Identical domain without identical morphemes is seen in sentence pairs of the form of *The letter is signed by him, The letter is signable by him*; we would seek some way of deriving both via an operator restricted to elementary sentences of the form *NtVN* ($V \neq be$).

As an example of considerations that affect the search for a derivation, we take the decision as to whether the insertion of adjective *A* before noun *N* and the insertion of *wh-* clauses after *N* should be taken as ϕ_a on *N* or as a product of certain ϕ on *S*, including the ϕ_c (connective) *wh-* (and its equivalents in other languages). We have, as incremental transforms of *A boy spoke up:*

> *A young boy spoke up.*
> *A young and hesitant boy spoke up.*
> *A boy who knew French spoke up.*

First, we note that these inserts can be at a distance from the *N* but not at a distance from the elementary sentence containing that *N*: A single adjective can thus be at a remove in some languages, as in Latin. For the distance restriction, note in English:

> *A boy spoke up, young and hesitant.*
> *A boy spoke up who knew French.*

This suggests that these inserts are to be defined as connected to the elementary sentence *A boy spoke up* (rather than to the *N*), since otherwise we could not state the restriction to their not being at a distance from the elementary sentence. Second, the acceptability of the resultant sentence depends (a) on the acceptability of the elementary sentence, and (b) on the acceptability not of the insert itself but of a sentence formed out of the insert plus the *N* in question. Thus *A cyclic boy spoke up* is unacceptable or almost so, as is *The boy is cyclic*, while *A cyclic flower was developed* is acceptable as is *The flower is cyclic*. Also *A boy who ridged on the northern shore spoke up* is unacceptable, as is *The boy ridged on the northern shore*, whereas *The rocks which ridged on the northern shore were of granite* is acceptable, as is *The rocks ridged on the northern shore*. We therefore derive these inserts from a second sentence, formed from the *N* in question plus *is* plus the insert, connected by *wh-* to the sentence that houses the insert.

Other examples of considerations which affect the search for a derivation may be seen in 4.2.2.1, 4, 5. A practical heuristic step in the case of a peculiarly formed set of sentences A is to seek some paraphrase set B of A such that B is a transform of A but is simpler in construction. We then seek the derivation of B, and try to find a product of base operators which will derive A from B.

Like transformations in general, the base operators act on propositional forms or on propositions. This presents certain inconveniences, since the immediately observable data of language are not propositions but sentences. However, we cannot define the operators as acting on sentences, because many sentences are ambiguous, and some operators act on one of the meanings and not the other; e.g., the passive acts on one meaning (one derivation) of *He marked two hours* (4.1.3). Since the different meanings of the ambiguous sentences result from different sequences of elementary sentences S_e and operators, we can define the operators as acting on S_e and on sentences which are defined as resultants of particular ϕ in the sequence. Thus when *He marked two hours* is characterized as S_e (disregarding the pronoun and number) we have a passive *Two hours were marked by him*, the passive (itself a product of base operators) being defined as operating on the elementary $NtVN$ form ($V \neq be$). However, *He marked two hours* is also characterized (as a different proposition) as the resultant of the following:

(S_e, of $NtVN$ form)	*He marked N_1*
($\phi_c \phi_c \dots \phi_c$)	$\rightarrow$ *He marked N_1 or He marked N_2 ... or He marked N_n*
(ϕ_z, type 2)	$\rightarrow$ *He marked N_1 or N_2 ... or N_2*
(ϕ_z, type 3)	$\rightarrow$ *He marked*
(ϕ_s)	$\rightarrow$ *His marking was for two hours*
($\phi_z \phi_c^{-1} \phi_s$)	$\rightarrow$ *He marked for two hours*
(ϕ_z on P)	$\rightarrow$ *He marked two hours*

On this resultant, the passive does not act, since the passive is not defined as operating on ϕ_z of P. (For the ϕ_c^{-1}, see 4.2.4.)

It is now possible to ask, in terms of the base transformations, how it is that there are ambiguous sentences: they are the phonemic degeneracies of transformational products. Each sentence of the language consists of a sentence S_e, or one or more of these with a particular ϕ-product on them. The resulting sentence is ambiguous if two or more different such derivations produce the same sequence of words or phonemes. This can happen: (a) if different word-classes, which are parts of different ϕ or S_e, have certain phonemically (or even morphemically) identical members and if these identical entities are brought into the sentence by their ϕ or S_e in such a way that they occupy the same position in the two resultants; (b) if different segments are zeroed in two resultants in such a way that the two residues of the resultants are phonemically identical. Thus ambiguity appears only when in certain resultants the traces of different S_e, ϕ (especially ϕ_z) are phonemically identical. An example of (a):

> *They ran up a bill* (pronounced from S_e, with *run up* $\in V$ and *bill* as its Ω)
> *They ran up a bill* (about some insects hastening on paper money:
> *They ran* pronounced from S_e, with *up* $\in P$ as part of a *PN* inner adjunct, see 4.3.1.)

An example of (b):

> *I left him little wiser* (one ϕ_z from *I left with him being little wiser.*)
> *I left him little wiser* (another ϕ_z from *I left him with me being little wiser.*)

The sentences in each pair are different transformational resultants with different meanings. Further transformations may operate differently on them, e.g., pronouning producing *They ran it up* from the first, but *They ran up it* from the second. It is only the morpheme or phoneme sequence that is the same, and this sequence can be transformationally interpreted at the cost of a limited degeneracy. Each phoneme sequence (sentence) can be obtained in only a few transformational derivations.

We therefore seek to define each operator as acting not on sentences, which are ambiguous, but on the unambiguous products of operators which characterize the particular grading memberships (grammatical meanings) of particular sentences.

4.2.3.2.   *Tables of products*

Furthermore, it turns out that it is not necessary to specify the whole products of operators which are the operands of a given ϕ_2; it is sufficient to say that ϕ_2 acts on all products which contain some particular ϕ_1, in a certain position, usually the last. This is seen as follows:

We first define each operation, whether a base operator or a fixed product of the analogic type (4.2.4) such as the passive, on its simplest operand: an elementary sentence (perhaps of particular form NtV, $NtVN$, etc.) in the case of the various subsets of ϕ_v, ϕ_s; a pair of residual sentences of particular form in the case of ϕ_c; the resultants of certain ϕ_s, ϕ_c in the case of ϕ_z; and the resultants of particular ϕ in the case of ϕ_p, ϕ_m. For convenience, we may consider the various forms of residual elementary sentence to be various subsets of a well-formedness (i.e., elementary sentence-making) operator ϕ_k. For each subset of each ϕ_2, we now note on what resultants of ϕ it acts, in addition to acting on the $S_e(=\phi_k)$ on which it was defined initially. Thus, many subsets of ϕ_v act on many or all subsets of ϕ_v and of ϕ_s:

> ... *is -ing* ... *on* ... *begin to* ...: *He is beginning to work.*
> ... *is -ing* ... *on* $S_n V_{s-} \Omega$: *His working is disturbing us.*

But *take a* ... *n* (member of ϕ_v), which acts only on a particular subclass of V of the residual sentences (*take a walk*, but $\nexists$ *take a lean*), is found to act on some ϕ_v and not on others: *He took a try at jumping*; $\nexists$ *He took a begin at jumping.*

For ϕ_2 to act on ϕ_1 means simply that the trace of ϕ_2 can be placed on the trace of ϕ_1, or on something present in all ϕ_1 resultants. And indeed we find that if ϕ_2 acts on the resultant of some ϕ_1, it will act on ϕ_1 no matter what operators had preceded ϕ_1 in the construction of the sentence: i.e., it will act on any sentence-characterizing sequence of operators ending in ϕ_1 (or in an unordered set of operators including ϕ_1, if the sequence of ϕ ends in such a set). This constitutes a major simplification in characterizing the set of sentences. For in any theory which composes sentences out of non-transformational elements (constituents, strings) only a hereditary stochastic process will suffice (3.2,6), whereas the composition of sentences from base transformations is a finite state process. This holds only with certain adjustments, especially that ϕ_m be treated as a set of variant phonemic spellings rather than as a transformation (since ϕ_m depends on two conditions, not one). And it requires that we accept among the ϕ_1 not only the base ϕ above but also certain fixed sequences of these ϕ, such as have to be set up for the analogic transformations (4.2.4).

The description of how a given ϕ_2, defined in the first place as acting on the elementary sentences, acts also on the resultants of certain ϕ_1 is made simpler by the fact that most resultants are similar to the elementary sentences, as word-class sequences. Thus, an elementary sentence is the word-class sequence $NtV\Omega$, where Ω is a family of word-class sequences; and the resultant of most base operators acting on $NtV\Omega$ is a similar sequence if we make the following evaluations: If the resultant of ϕ_a produces, say, $NtxV\Omega$, where x is a modifier of V, we consider $xV \approx V$ (read: xV fills the position of V in the sentence form). When ϕ_v produces $NtV_v Va \Omega$, we consider $V_v \in \{V\}$ (read: V_v is a member of the class of V), and $Va \Omega$ to be the new Ω of V_v. When ϕ_s produces $Sn t V_{s-} \Omega$ (e.g., *His coming surprised me*) we consider $Sn \approx N$; when ϕ_s produces $NtV_{-s} Sn$, we consider $Sn \approx \Omega$. The material added by ϕ_c (namely CS_2 and its permuted, adverbial, and zeroed forms), and the adverbial and ϕ_c-like forms of ϕ_s (see 4.2.2.1), can be considered like the material added by ϕ_a. ϕ_z rarely presents problems, because the NtV of S_1 are not generally zeroed.[24] ϕ_p, however, produces new forms, as in *This I like*. When a base operator acts not on an elementary sentence but on a resultant of a base operator, or when two base operators act simultaneously (unordered), the above process repeats.

In accordance with this similarity of resultants as sequences of word classes, many resultants of one base operator have forms which are similar, in the sense of $\approx$ above, to the resultants of other base operators. ϕ_s have ϕ_a-like forms; ϕ_c have ϕ_s-like and ϕ_a-like forms. Many ϕ_s have ϕ_v-like forms. The change of one resultant to the form of another must itself be produced by some base operators, usually ϕ_z (as in 4.2.2.1–4).

The similarity in resultants does not destroy the separate recognizability of the trace of each operator. For the similarity is in the superclasses N, V, whereas the trace of an operator is in the particular subclass of N, V, etc., or in the particular sentence deformation or zeroing which is brought in by the particular operator.

It is therefore simple to extend the original definition of each ϕ so as to include its other operands than S_e. Given the original definition of ϕ_v: $NtV\Omega \rightarrow NtV_v\,Va/n\Omega$, it follows that ϕ_v on the resultant of ϕ_v (i.e., $\phi_v\phi_v$): $NtV_v\,Va/n\Omega \rightarrow NtV_v\,V_v a/nVa/n\Omega$ (*He began trying to study English*), etc. In the case of ϕ_c, the S_1 on which it operates does not change, and the CS_2 which is introduced by it is simply an adjunction to the S_1; hence, for example, ϕ_v on a resultant of ϕ_c: $NtV\Omega CS \rightarrow NtV_v\,Va/n\Omega CS$ (*He writes to make money $\rightarrow$ He stopped writing to make money*; i.e., He now writes as he pleases).[25]

Given the operand, there is a 1–1 correspondence between ϕ_1 and the resultant of ϕ_1. We can therefore define the ϕ as operating on ϕ, rather than on resultants of ϕ, namely propositions. There are advantages to this. One is that this permits a clearer formulation of the relations within the set of ϕ. Another is that there exist exceptional cases (analogic and

---

[24] In certain aberrant cases, given a pronoun (4.2.2.6:3) with a *wh* adjunct (or adjective derived therefrom), the pronoun is zeroed; but the original structure is salvaged because the remaining adjunct takes its syntactic place, i.e., $\approx N$. E.g., *The large is better*. ← *The large one is better.*; *I heard what he said.* ← *I heard that which he said.*

[25] We have here an ambiguous sentence, for the resultant proposition is the same word sequence as the resultant of ϕ_c on ϕ_v: $NtV_vVa/n\Omega \rightarrow NtV_vVa/n\Omega CS$ (*He stopped writing $\rightarrow$ He stopped writing, to make money;* He stopped in order to make money). In this case, a comma is optional.

regularizing operations) in which the operand is not an occurring sentence, although very similar to a sentence (4.2.4, and Chapter 6). The chief advantage is that we can state the operands of ϕ_2 in terms of transitions in a ϕ table, rather than in terms of long lists of sentence forms. In constructing such a table, it is necessary to consider all the ϕ in a particular ϕ product as acting on the same elementary sentence. This requires that ϕ_c be treated not as a binary operator on two sentences but as a unary operator on S_1 that conjoins CS_2 to S_1 (4.3.2).

The characterization of propositions as successions of ϕ is thus seen to have the Markov property. If we think in terms of all inter-ϕ transitions which have a positive probability of being part of a proposition characterization as distinct from those that do not, then the ϕ sequences which characterize propositions are a Markov chain.

Within the set ϕ, each of the types of operator has several subsets, some of which act only on a subset of the operand, and whose effects upon their operand differ in detail. Therefore a detailed study of how each subset of the base types operates on every other subset is no simple matter. However, it is easy to make a table showing which type of operators can act on which, omitting details:

	ϕ_k	ϕ_v	ϕ_s	ϕ_c	ϕ_p	ϕ_z
ϕ_v	+.	+.	+.	+.		−
ϕ_s	+	+	+	+	+.	+.
ϕ_c	+	+	+	+.	+.	+.
ϕ_p	−		+.	+.	+.	+.
ϕ_z	−	−	+.	+.	−	

+   indicates that all operators of the row act on all of the column.
+. indicates that all or some of the row act on some of the column, in accordance with the definitions of the subsets of operators.
−   indicates that none do. Absence of mark indicates that the operators of the row and column act independently of each other on a common operand: e.g.,

ϕ_p: *I choose this → This I choose*, ϕ_v: *I choose this →*
I have chosen this; together they yield *This I have chosen.*

The table is obtained purely from the matching of the arguments and resultants of each operation: If the resultant of ϕ_1 is the same sequence of N, V, Ω, and adjunct forms as the argument defined for ϕ_2, and if in this sequence the subclasses to which the argument of ϕ_2 is restricted (if any) are included in the subclasses in the resultant of ϕ_1, then ϕ_2 can act on ϕ_1. The transformational derivation of a sentence thus depends on matching the subclasses, as implicit in the definition of $\rightarrow$ (4.1.5.2). No further ordering of the operators need be imposed in order to account for all existing products of base operators. And as we have seen in the independent cases above (and in the other commutative cases), some pairs of ϕ are not ordered among themselves.

4.2.4. Analogic products

Up to this point no cognizance has been taken of the fact that all languages change (2.6). We disregard here the cumulative changes in sound and hence occasionally in phonemic distinctions, the changes in word meaning, the borrowing and innovation of words and occasionally of syntactic sequences which are in most cases fitted into the existing syntactic system. We consider only the change of the syntactic system: of the domains (and, rarely, the traces) of transformations; of the subclasses defined as transformational domains; and of the sentence forms (or segments of them) defined as transformational resultants. Any detailed survey of the transformations of a language reveals several which are obviously in process of formation or change. We note here an irregularity of transformations which affects the statement of how transformations operate, and which is one of the contributors to the development of transformations.

The base operators above were obtained by considering the morphemic trace of each transformational equivalence relation between sentences, and trying to find morphemic components of which that trace could be a product. The intention was that each of these base components should itself constitute the trace of a transformational relation between sentences. So that if *I know that he began to speak* is the $\phi_s \phi_v$ transform of *He spoke*, it is also the ϕ_s transform of a sentence *He began to speak*, which is the ϕ_v transform of a sentence *He spoke*. In the case of certain operands, however, the intermediate products do not exist as sentences, so that the base trace-producing operators in these cases are not transformations although their product is a transformation.

A simple case of this is *is going to*, a member of ϕ_v (e.g., *He is going to stay here*). This is similar in form to *is ... ing*, a member of ϕ_v (*He smokes*

$\rightarrow$ *He is smoking*), operating on *go to*;[26] but *go to* is not found as a member of ϕ_v ($\not\exists$ *He goes to stay here*, except as a case of ϕ_c, transform of *He goes in order to stay here*). And this *go to* cannot be the *V* of an elementary sentence, for the follower there, the Ω, would be a noun and not a verb. Hence we have to say that this *go to* is indeed a member of ϕ_v, into which it indeed fits syntactically, but that only the product of it with another ϕ_v, *is ... ing*, occurs as a sentence.

Somewhat similarly, *He gave a party, He had a party, He made a party*, etc. looks like a resultant of ϕ_v (as in ϕ_v: *He looked* $\rightarrow$ *He gave a look, He had a look, He took a look*); but $\not\exists$ *He partied*. We say that in this case only the product $\phi_v \phi_k$ occurs, and not the resultant of ϕ_k alone.

Once we grant that components of an operator product may have unacceptable *S* as their intermediate resultant, many problematic transformations are found to be products of the base operators. To take a peculiar case: all question forms in English are obtained by a single chain of base operators. For example, when the question asks for the object of the verb we find:

$\quad\quad\quad$ (1) *What will he write?* (Ans.: *A letter*)
(by ϕ_z) $\leftarrow$ (2) *I ask: what will he write?*
($\phi_m \phi_p$) $\leftarrow$ (3) *I ask what he will write.*
($\phi_z \phi_p$) $\leftarrow$ (4) *I ask whether he will write N_1 or N_2 ... or a letter ... or N_n.*
(ϕ_z) $\quad\leftarrow$ (5) *I ask whether he will write N_1 ... or he will write N_n.*
(ϕ_s) $\quad\leftarrow$ (6) *He will write N_1 ... He will write N_n.*[27]

If the question asks for the verb, we find:

$\quad\quad\quad$ *What will he do?* (Ans. *Smoke*). This should be obtained
(by ϕ_z) $\leftarrow$ *I ask: what will he do?*
($\phi_m \phi_p$) $\leftarrow$ *I ask what he will do.*
($\phi_z \phi_p$) $\leftarrow$ *I ask whether he will do V_1ing or V_2ing ... or smoking ... or $V_n$ing.*
(ϕ_z) $\quad\leftarrow$ *I ask whether he will do V_1ing ... or he will do $V_n$ing.*
(ϕ_s) $\quad\leftarrow$ *He will do V_1ing ... He will do smoking ... He will do $V_n$ing.*
(ϕ_v) $\quad\leftarrow$ *He will V_1 ... He will smoke ... He will V_n.*

[26] If *is going to* belongs to ϕ_v and is composed of *is ... ing* belonging to ϕ_v operating on *go to*, then *go to* could belong to ϕ_v; for the product of two members of ϕ_v is in ϕ_v.

[27] Note that the operators between (1) and (5) are all paraphrastic, so that (1) is a paraphrase of (5), although a transform also of (6). ϕ_z : (5) $\rightarrow$ (4) zeroes the repeated *he will write*; ϕ_p permutes $N_1 ...$ *or N_n* to after *whether*, and ϕ_z automatically pronouns this disjunction to *-at* and combines it (automatic ϕ_m) with *whether*, to yield *what*, thus producing (3). In (3) $\rightarrow$ (2), ϕ_m introduces the intonation.

But $\nexists$ *He does smoking*, although the ϕ_v occurs for occupational verbs as in *He does writing*, etc. We must now say that *do ... ing* as a member of ϕ_v operates also on nonoccupational verbs, but only as a component of a product, e.g., with *I ask* and with the pronouning of *whether V_1ing or V_2ing ... or $V_n$ing to what*.[28]

The characterization of these marginal cases can be brought out with the aid of the following example. Sentences like

> *These shelters build easily.*
> *The play read beautifully.*

which change object into subject and unaccountably require an adverb (and apparently a particular subclass of adverb at that), can be explained in somewhat the following manner:

N_1 *read the play.*

$(\phi_c) \quad \rightarrow \quad N_1$ *read the play or N_2 read the play ... or N_n read the play.*

$(\phi_z) \quad \rightarrow \quad N_1$ *read the play, or N_2 ... or N_n.*

$(\phi_p) \quad \rightarrow \quad N_1$ *or N_2 ... or N_n read the play.*

$(\phi_s) \quad \rightarrow \quad$ *The reading of the play by N_1 or N_2 ... or N_n was beautiful,* (or perhaps: *went beautifully*).

$(\phi_z) \quad \rightarrow \quad$ (1') *The reading of the play was beautiful.*

$(\phi_z \phi_s) \rightarrow \quad$ (3) *The play read beautifully.*

The subclass of adjective of manner which is introduced in ϕ_s is apparently a particular one which goes with indefinite subject [the disjunction of N which is zeroed in (1')]. To understand the last step, we consider the common *The Ving of Ω by N is A or N's Ving of Ω is A $\leftrightarrow$ NtVΩ Aly*: e.g., *His reading of the play was hurried $\leftrightarrow$ He read the play hurriedly.* There is still some question as to what succession of base operators would derive one of these from the other, but the transformational relation is unquestionable. In the case of Ω = zero, we have

(2) *The Ving of N is A or N's Ving is A $\leftrightarrow$ NtV Aly*

The singing of the birds was beautiful $\leftrightarrow$ The birds sang beautifully.

In the preceding example, the form of (1') is (1) *The Ving of N is A*, where N is the original Ω; the original subject, *by N*, has been zeroed. As word-class sequence, this (1) is identical with (2), where the N was originally the subject and not the object of the verb. If then we apply to (1) the

---

[28] In *He does a lot of smoking, He does some smoking*, we have the product of ϕ_v (*do ... ing*) and ϕ_a (*a lot, some*), but still this ϕ_v does not occur by itself on *smoke*.

same transformation which operates above on (2), we obtain

(3) *The play read beautifully.*

The odd effect in (3) is due to our operating on the object as though it was the subject, something which was possible only because (1) and (2) were identical word-class sequences.

What has happened here is that a product of transformations which related two sentence forms $A \leftrightarrow B$ has been extended to a sentence form A' to obtain B'. A' is the same sequence of word classes as is A, but has not the same word subclasses. In (1') and (1) the *of N* had the subclass restrictions of the object of the preceding *Ving* (i.e., the N preceding the V in the $NtV\Omega$ from which (1') was obtained by ϕ_s), while in (2) the *of N* had the subclass restrictions of the subject of the preceding *Ving*. Extension of a product of operators from one subset of values of a variable to another of the same variable suffices to describe all the cases of this transformational type, including the preceding examples. E.g., in the preceding example, the relation *He will write* $\leftrightarrow$ *What will he do?* which can be obtained via *He will do writing*, which occurs as a sentence, is extended to *He will smoke*, where the intermediate *He will do smoking* does not naturally occur.

In some cases, the operator which is extended to A' is the inverse of an operator which acts on A (the inverse of a transformation being still a transformation). If we consider a whole family of sentence rearrangements such as

His purchase was of prints.
The purchase of prints was by him.

we find that we can obtain them, perhaps in more than one way, by means of the base operators, but only if some are taken in the inverse sense. E.g.

		He purchased prints
(ϕ_s)	$\to$ (1)	*The purchase of prints by him took place*
(ϕ_z^{-1})	$\to$	*The purchase of prints which was by him took place*
(ϕ_c^{-1})	$\to$ (2)	$\Big\{$ *The purchase of prints took place.*
		The purchase of prints was by him.

In (1), which results by ϕ_s, we have a word-class sequence N of N by NtV which is identical with that of (3) below, which results by ϕ_z on ϕ_c:

	(4)	$\Big\{$ *The folio of prints fell.*
		The folio of prints was by the edge.
(ϕ_c)	$\to$	*The folio of prints which was by the edge fell.*
(ϕ_z)	$\to$ (3)	*The folio of prints by the edge fell.*

Applying to (1) the inverse, $(\phi_z \phi_c)^{-1}$, of the ordered operators which had produced (3) out of (4), we obtain (2), which is a pair of sentences having the same word-class sequences as in (4): *N of NtV. N of Nt be by N.* The subclasses of words are not the same: *purchase* is a *Vn* (verbs with nominalizing suffix); *by* in (1) is a separate subclass of *P* which appears as part of ϕ_s, but *by* in (3) is a member of a large subclass of *P*; *him* is a member of whatever subclass appears as *N* in *N purchased* Ω; *took place* is a member of ϕ_s, and is in its zeroable subclass ϕ_{s_0} (*took place, occurred,* 5.2), for which ϕ_z: $\phi_{s_0}Sn \rightarrow S$ (*His arrival took place* $\rightarrow$ *He arrived*), a necessary step toward the later dropping of the first sentence of (2). The *N of Nt be by N* form of (4), which is derived from elementary sentence forms defined on a domain of concrete-word subclasses, would not accept the subclasses found in (2) without an extension of domain on the basis of gross word-class similarity and inverse transformations.

In all these cases the claim is not that the sentences in question, the final ones of each derivation, are derived historically or even structurally by the given path of operations. It is only claimed that these sentences, which are indeed transforms of the initial sentences of each derivation, differ from the initial sentences only by a product of elementary traces (or inverses of traces). If each step in the derivation consists of an occurring sentence (or a formula for one), these cases differ from the ordinary representation of transformations, as products of elementary operators, by the fact that some of the individual steps in the derivation consist of inverses or whole products of elementary operators. In all such cases the product of operators has been extended, from some subclass on which the operators have been defined as independent transformations, to a similar subclass on which the operators have not been defined (e.g., for morphological reasons, or because the operator is defined in the opposite sense).

We can characterize all these cases by saying that if a transformational relation exists between two sentence forms $A(x_1)$, $B(x_1)$ which contain some subclass x_1 of a word class x, then given the same sentence form $A(x_2)$ of a similar subclass x_2 of x, it is possible to obtain $B(x_2)$:

$$A(x_1) \leftrightarrow B(x_1)$$
$$\underline{A(x_2)}$$
$$B(x_2)$$

The acceptability of $B(x_2)$ depends on the similarities between x_1 and x_2 in ways that have not yet been determined, and that may lie beyond the scope of purely structural analysis. To put it in other words:

$$\phi_j \ldots \phi_k: \quad A(x_1) \rightarrow B(x_1)$$

implies some probability, for x_1 similar to x_2, of

$$\phi_j \ldots \phi_k: \qquad A(x_2) \to B(x_2)$$
$$(\phi_j \ldots \phi_k)^{-1}: \qquad B(x_2) \to A(x_2).$$

4.2.5. Summary

Starting with transformational traces as a large set of differences among transformationally related graded sentences, we found that many of these could be obtained as products in a small set of base traces (i.e., sentence differences), each of which could be considered as due to a directed base operator. For any base operators ϕ_i, ϕ_j, the product $\phi_j \phi_i$ (ϕ_j acting on ϕ_i) exists provided that there is a nonempty intersection of the range of ϕ_i and the domain of ϕ_j. The remaining transformational traces (4.2.4), which do not satisfy this condition, turn out to be a small selection from among the products of base operators which could be obtained if we relaxed the requirement that the subclasses specified in the intersection of range and domain be the same, and if in some cases we removed the directedness of the operator (i.e., accepted inverse operators). All transformational traces are thus seen to be sums of the base sentence differences —the regular ones (4.2.3) being all those which satisfy subclass matching and directedness, and the analogic ones (4.2.4) being a selection of those which relax these demands.

We now have a set of intersentence differences, which are products of base relations, where each base relation is either one of the base ϕ operators defined on particular subclasses of words, or else one of certain fixed successions of the base operators or their inverses applied analogically to different subclasses than those defined for the component ϕ occurring independently. More precisely, each ϕ_j is defined as operating on particular sentence forms (in most cases elementary) of particular word subclasses, and also on particular ϕ_i (i.e., on the resultant of ϕ_i operating on a sentence). In the product $\phi_j \phi_i$, the set of the resultants of ϕ_i (the resultant being a form with certain variables) must include the domain of the argument on which ϕ_j is defined.

One significance of this analysis is that the characterization of sentences requires no other operation among transformations than this multiplication. Hence if ϕ_i is involved in a sentence, it must be a member of the succession of ϕ which yield that resultant. Thus in sentences of the form *The play read beautifully*, in which (except for a few idiomatic cases of zeroing)[29] an adverb is always present, we saw that the adverb was a necessary step in the derivation.

---

[29] We can permit such exceptions, because they are due to stateable cases of later operations.

These base relations satisfy the empirical conditions for the linguistic transformations described above, relating to the preservation of the order of acceptability of sentences. Each of these relations has a single meaning effect which it contributes to all sentences on which it operates. The incremental ϕ_a, ϕ_v, ϕ_s, ϕ_c generally change the meaning, but ϕ_p and ϕ_z (together with pronouning) and ϕ_m do not (i.e., are paraphrastic), although ϕ_z can have an indirect effect on the meaning of a sentence in that it can introduce degeneracies in form, yielding ambiguous sentences. Also paraphrastic are the fixed (analogic) ϕ sequences which consist of ϕ_s, ϕ_c and their zeroes or inverses (but which are such as to leave a changed form of the sentence), e.g., the passive, or *He purchased these books* → *His purchase was of these books.*

The trace of each ϕ in the sentence is the introduction of a constant, or a permutation or zeroing (or pro-wording) of constants or of word classes. Because resultants often preserve the gross word-class sequence of residual (elementary) sentence forms, the trace of ϕ_z or ϕ_p is often recognized not by different word classes but by a different subclass (which is due to the ϕ_z or ϕ_p) appearing in the position where another sentence form has the same class but not the same subclass (e.g., *He spoke two hours, He spoke two words*). Thus traces are never entirely lost, even traces of ϕ_z. The location in the sentence of the trace of each ϕ_j can be determined from the sequence of ϕ which represents the sentence; we include here in the set of ϕ the elementary-sentence-making ϕ_k introduced in 4.2.3. The trace of each ϕ_j is located in the ϕ_i on which it operates (i.e., which is its argument), or if ϕ_i does not contain the word classes which constitute the argument of ϕ_j then the trace of ϕ_j is located in the nearest ϕ (in the sequence of ϕ) which does contain the argument of ϕ_j. Thus in $\phi_v \phi_s \phi_k$ the trace of ϕ_v is in ϕ_s (*I began to report his arrival*); in $\phi_s \phi_v \phi_k$ the trace of ϕ_s is partly on the ϕ_v and partly on the ϕ_k (*Their beginning to arrive aroused us*); and in $\phi_v \phi_c \phi_k$ the trace of ϕ_v is entirely in ϕ_k but it acts after ϕ_c has acted on ϕ_k (*He stopped writing to make money*, i.e., stopped money-oriented writing, see fn. 25).[30]

Because of the inverse operators which occur in analogic products of operators, there are cases in which a particular word subclass is not uniquely the trace of a particular ϕ. For example, not all occurrences of C are traces of ϕ_c: some are traces of ϕ_m^{-1} analogic to $\phi_m \phi_c$. Thus on the

---

[30] In defining the individual ϕ one can seek to maximize the cases in which the trace of a ϕ is located in the ϕ on which it operates. Thus if ϕ_v is defined as a special case of ϕ_s (for the case where the subject of both the ϕ and the S must be the same, then the above ϕ_v in *They began to arrive* (now ← *They began their arrival*) contains the trace of the ϕ_s in *Their beginning to arrive aroused us.*

analogy of:

> *He smiled. He entered*
>
> (ϕ_c) → *He was smiling when he entered*
>
> (ϕ_m) → *He entered (while) smiling; He entered smilingly.*

we could analyze

> *He writes Russian illegibly*
>
> (ϕ_m^{-1}) → *He is illegible when he writes Russian.*

The *when* in the last sentence is thus not due to the operation of any ϕ_c on the component sentences.

It remains to mention that we have in transformational analysis a single theory accounting for virtually all syntactic phenomena, i.e., all phenomena after the dictionary, which lists particular phoneme sequences as being members of particular morpheme classes. Not only do all sentences of the language participate in (ultimately base) transformational relations, but they all are decomposable by base transformations into elementary sentences. The great bulk of the changes which come upon words result from transformations on the sentences containing those words. Thus most of morphology (the addition of suffixes, etc.) is simply a particular word's share in a transformational trace: e.g., the morphological change from *analyze* to *analysis* is simply a section of the transformational trace in ϕ_v: *He analyzed it → He made an analysis of it*. Similarly, a shift in word-meaning is the deposit of the transformations on the sentence containing that word: e.g., the difference in meaning between *to talk* and *a talk* is part of the meaning brought in by ϕ_v: *He talked about it → He gave a talk about it*. So also the two meanings of *speech* arise when that word appears in two different transformational resultants: ϕ_v: *He speaks → He makes a speech*, ϕ_s (of manner): *He speaks → His manner of speaking is labored, His speech is labored*. Furthermore, many apparently special facts about sentence structure turn out to be merely special cases of the regular operation of transformations: such an explanation, for example, can account for the special order of adjectives before the noun (e.g. *large white box* rather than *white large box*).

4.3. Sets of operators and of sentences

In characterizing sentences in terms of transformational operators, we can define certain relevant sets of objects whose properties can be investigated (4.3), and from which we can define subsets, or differently organized sets, of interest to sentence characterization (Chapter 5).

4.3.1. *The set of transformations under multiplication*

The base operators $\phi: A(x) \to B(x)$, where A, B are propositional forms and x a particular word class or subclass or ϕ trace present in $A(x)$, can be taken as generators of the set of transformational relations among sentences.

In considering the set of ϕ products, it is possible to obtain a stronger result than the statement (4.2.5) that $\phi_i \phi_i$ exists if the intersection of the domain of ϕ_j and the range of ϕ_i is not empty. There is an important subset ψ of transformations, including most non-paraphrastic ϕ, whose domain is the whole subset of sentences S_ψ which contain no ϕ from the complement set of ψ; i.e. the sentences of S_ψ are formed only out of S_e and ψ.

To see this, we note that most ϕ_a do not repeat and do not occur freely on other ϕ_a (e.g., *some*). Certain subsets of ϕ_v do not repeat (e.g., *have* in *I had a walk*), or do not occur on all other ϕ_v or on all S_e (e.g., *is ... ing* does not occur on *He knows it*); but other subsets, $\psi_v \subset \phi_v$ occur on all ϕ, S_e (e.g. *begin to* in *I began to begin to work*). Certain subsets of ϕ_s (e.g., ones taking deformations 4 and 5 of their operand, as in *frequent, slow*) do not repeat ($\nexists$ *His going slowly was slow*), but the other subsets, $\psi_s \subset \phi_s$, do (as in *That he came is false is false*) and also occur on all ϕ, S_e. As to ϕ_c, it seems that all products of them occur, except for *wh* which does not operate on ϕ_c; however, we can salvage the closure of ϕ_c under multiplication by saying that *wh* $\phi_c = \phi_c$ *wh*. Furthermore, most ϕ_c do not occur on arbitrary S-pairs; but we can treat each ϕ_c as a unary operation concatenating particular secondary S to arbitrary primary S, or we can say that each ϕ_c operates on arbitrary pairs S_1, S_2 provided the resultant is taken not as $S_1 C S_2$ but as $S_1 C S_2$ followed by particular $CS...CS$ of a kind determined by the ϕ_c. Finally we note as a matter of empirical fact that if ϕ_g, ϕ_h are subsets in each of which every member combines with every member of the subset, then each member of ϕ_g also combines with each member of ϕ_h.

As to the paraphrastic ϕ_p, ϕ_z, ϕ_m, they require very special conditions in their operand. They occur only in particular combinations with other ϕ, and never directly on S_e. Hence these are not in ψ.

It follows from all this that if we consider ψ, the set of all members of $\psi_v \subset \phi_v$, $\psi_s \subset \phi_s$, ϕ_c, and their products, then each member of the set, operating on arbitrary S, represents a sentence; and the set is closed with respect to multiplication,[31] and hence unbounded (since there is no general inverse).[32] Of course, in the actual language the products of transformations in any one sentence are quite short, but given an arbitrary product of length n one can always operate with some ϕ to obtain a product of length $n+1$.

The products of base operators are associative: e.g., $\phi_s(\phi_s \phi_v) = (\phi_s \phi_s)\phi_v$; i.e., both produce the same proposition, as in *I know that he thinks she has come*. This is one of the great differences between representing a sentence as a sequence of words and representing it as a sequence of transformational operators (4.3.2). The word sequence of a sentence is nonassociative, and different groupings of the same words may have different meanings (as in footnote 25). But each operator product represents only one grammatical meaning of a sentence, i.e., only one proposition.

The associativity of ϕ_c may seem strange, since binary C is not associative: $S_1 C(S_2 \, CS_3) \neq (S_1 \, CS_2)CS_3$, i.e., these do not produce the same proposition; an example of the first is *He left because* (*she had insulted him after everybody arrived*), and of the second (*He left because she had insulted him*) *after everybody arrived*. (The parentheses in the sentences are to indicate the two meanings.) However, when ϕ_c is taken as a unary on S_1, concatenating CS_2 to it,

then $\phi_{c2}: S_1 \rightarrow S_1 CS_2$,
         $\phi_{c3}: (S_1 CS_2) \rightarrow (S_1 CS_2)\, CS_3$;

thus $S_1 CS_2 = \phi_{c2} \, S_1$; $(S_1 CS_2)\, CS_3 = \phi_{c3} \, \phi_{c2} \, S_1$.

But $\phi_{c3}: S_2 \rightarrow S_2 \, CS_3$,
         $\phi_{c\phi_{c3}s_2}: S_1 \rightarrow S_1 C(S_2 \, CS_3)$;

thus $S_2 \, CS_3 = \phi_{c3} \, S_2$; $S_1 C(S_2 \, CS_3) = \phi_{c\phi_{c3}s_2} \, S_1$.

The two different meanings of the sentence above are thus produced by different ϕ_c operators. As to the products of ϕ_c, they are associative:

[31] Various ways may be attempted for reducing the complement set (ϕ-ψ) of ψ. One can seek synonyms in ψ for members of (ϕ-ψ), as in 6.5. One can include ϕ_{com}, from the complement set, in ψ by setting $\phi_{com} \, \psi_i = \psi_i$ in the case where the resultant of ψ_i is not in the domain of ϕ_{com}, but this will give a false picture of the composition of the sentence which is represented thereby. The analogic transformations can be included as single ϕ, instead of irregular products of base ϕ, as follows: Whenever an analogic fixed sequence $\phi_a \ldots \phi_e$ of base operators (or their inverses) acts on $A(x')$, where x' is different from the word-subclass on which ϕ_e is defined, we define $\phi_{a \ldots e}$ as an operator on $A(x')$, with trace $\phi_{a \ldots e} = $ trace $\phi_a \ldots \phi_e$. But they remain in ϕ-ψ.

[32] We can go further and define the inverse ϕ_i^{-1} of ϕ_i to be the removal of the trace of ϕ_i. If certain subclass distinctions are disregarded, a sentence A may have a word sequence identical with the trace of ϕ_i even if ϕ_i has not operated in A (4.2.4). Then ϕ_i^{-1} may operate on A (i.e., on the ϕ sequence which produced A), as well as on ϕ_i. For all other ϕ_j, we would have $\phi_i^{-1}\phi_j = \phi_j$, as above.

$\phi_{c4}(\phi_{c3}\,\phi_{c2}) = (\phi_{c4}\,\phi_{c3})\phi_{c2}$; both produce the same grouping (association) of C and S, namely in this case $((S_1CS_2)CS_3)CS_4$.

The products are in general not commutative; e.g., $\phi_v\,\phi_s$ (*I began to believe that he left*) $\neq \phi_s\,\phi_v$ (*I believe that he began to leave*).

In the case of the mutually independent operators (with no mark in the table above) we may find it convenient to say not only that they are commutative, but alternatively that they do not act on each other: *This I have chosen* is the resultant of unordered ϕ_p and ϕ_v. This situation also holds for $\phi_z\,\phi_z$. And it holds for one subset of ϕ_c: If a sentence S_1 receives two *wh-S* which begin with different N of S_1, the two *wh-S* operate independently of each other on S_1:

> *wh: A man$_1$ saw a man$_2$, The man$_1$ is young* → *A young man saw a man;*
>
> *wh: A man$_1$ saw a man$_2$, The man$_2$ is old* → *A man saw an old man;* together they yield *A young man saw an old man.*[33]

The products of the ψ then form a semigroup, or monoid if we define $\phi_1 = 1$ as the identity transformation.

The transformations in ϕ-ψ, whose products are restricted, fall into a few types which differ from ψ in various respects. All of them do not occur freely on transformations of their own subset, even if they otherwise occur on all sentences (i.e., on all transformations which produce those sentences): the ϕ_a quantifiers, the ϕ_v of time (e.g., *have gone*), the ϕ_s of occurrence (e.g., *on Tuesday*). In addition, some of them do not occur on all S_e types: certain ϕ_v of mode (e.g., *It was very moving* ← *It moved us*); and the ϕ_s of manner (e.g., *is slow*) which do not occur on sentences whose V is *is*; furthermore, particular choices of these adverbs of manner are more acceptable with particular V choices in their operand sentence, so that they are like extensions of the V rather than operators on it. All of these ϕ-ψ can be considered an inner set of ϕ, closer to the operand than are the ψ. Finally, the paraphrastic transformations differ from all others, in being restricted to particular operand conditions (never S_e) and in adding no semantic contribution. They could be considered an outer set of ϕ.

4.3.2. *The set of sentences under transformations*

The ϕ are partial transformations in the set S: each ϕ maps a subset of S into S. In the important subset S_ψ of S, the subset ψ of ϕ are transfor-

---

[33] The subscripts are only to indicate the distinguishing of the first and second N of S_1, which can be done in the grammar by the method of 5.6.3, 5.7. The fact that *wh*-individuates the first N of each S_2 is derived in the further analysis of ϕ_c (5.7); the occurrence of *a* and *the* here is related to this.

mations: each maps the whole of S_ψ into S_ψ. In the set S as a whole there is perhaps no ϕ which can operate on all members of the set (these include questions and other resultants of ϕ_z and ϕ_m). However, for every member of S there are some ϕ which operate on it to produce a member of S. And given the set ϕ defined in a suitable way, every member of S has a decomposition by members of the set ϕ into a particular (partially ordered) subset of residual sentences (or rather, of occurrences of these, since some may appear more than once in the decompositions).

The addition of a ϕ to a sentence does not alter the ϕ composition of the operand. In particular, the S_e and incremental ϕ (i.e., ϕ_a, ϕ_v, ϕ_s, ϕ_c) composition of a sentence, and the meaning which these determine for the sentence, are invariant under all nonincremental ϕ (ϕ_p, ϕ_z, ϕ_m, analogic ϕ). We could consider the S_e and incremental ϕ to be the constructors of a meaning-bearing proposition, and the other ϕ to produce deformational transforms of that proposition.

Different structurings of the set of sentences arise from different formulations of the operators.

We can define ϕ_c as binary operators: $S, S \to S$, and all the other ϕ (including the analogic fixed sequences of ϕ) as unary: $S \to S$. If a unary ϕ_j is defined on the whole set S as domain, then $\phi_j : S \to S$ is a 1–1 mapping of the set of propositions into itself, while the other unaries are isomorphic partial transformations in the set S, each ϕ_i sending S of one subset (which satisfy the conditions for the argument of the ϕ_i) onto another subset. As in 4.3.1, they could be considered transformations on the whole set S, if we set $\phi_a S_b = 1.S_b$ (i.e., S_b is left unchanged) when S_b does not satisfy the conditions for an argument of ϕ_a. (But this would destroy the uniqueness of decomposition for S_b.) The interest in such a formulation would depend in part on the skew effect within the set S of different portions of the set being left unchanged as various transformations are successively carried out over the whole set. Such extension of partial transformations can safely be made for many paraphrastic base operators, ϕ_p, ϕ_z, ϕ_m (e.g., ϕ_z: *I went* $\to$ *I went*).

As to ϕ_c: a subset of ϕ_c acts as a binary composition in the set S. For example, *and*, and to a lesser extent *or*, may be found (with somewhat different effects in some cases) between any two S, even between question and assertion, etc. (*Will you go, or else I will go*; *I will go, and will you go?*). The remaining ϕ_c are partial transformations from a subset of S pairs to a subset of S. If we wish to define another subset of ϕ_c, namely C_s, as operating on all S pairs, we will find (5.6) that an arbitrary $C_s S_2$ may be added to S_1 only at the cost of adding certain additional CS (to adjust to the required similarities and differences between S_2 and S_1). Therefore, the resultant of a binary operation C_s on arbitrary S_1, S_2 is not simply $S_1 C_s S_2$ but $S_1 C_s S_2 \, CS \ldots CS$, with various of the added CS being zero-able in various cases.[34]

The ϕ_c can also be considered a unary operator. This is possible because in ϕ_c: S_1, $S_2 \rightarrow S_3$, the S_1 remains unchanged within S_3 (except in certain cases where later operators act), and because the S_2 depends in part upon S_1 (in respect to word choice), especially if the resultant is required to be $S_1 C S_2$, with all the further $CS \ldots CS$ zeroable. We can therefore define those conjunctional cases whose resultant is simple SCS as ϕ_{ck}: $S_1 \rightarrow S_1 C_k S_{k1}$ where S_{k1} is any sentence of a family of sentences determined by S_1 and the particular C_k.

Given a suitable list of base operators in a language, products of which can be formed as in 4.2.3, each graded sentence (proposition) in the language can be characterized by a unique product of elementary sentences and base operators, aside from commutativity of certain operator products. If two operator products which differ in more than this local commutativity produce the same word sequence, then the word sequence is an ambiguous sentence, i.e., it represents two graded sentences. The only case in which a single graded sentence (a nonambiguous sentence, or a single reading of an ambiguous one) can be reached by more than one product of S_e and operators arises when the product contains an analogic transformation for which alternative ways exist for reducing it to base operators (4.2.4). If the analogic transformations (e.g., the passive) are considered as single operators in addition to the base operators, or if each of them is given a preferred decomposition to the exclusion of alternatives, then each pro-position is a unique product (aside from local commutativity).

The special provision for commutative subsequences of an operator sequence that characterizes a graded sentence can be obviated if we say that what characterizes a graded sentence is not a sequence of operators but a partially ordered set of operators, which is identical with the sequence except that each commutative subsequence is an unordered set at the point in the sequence that was occupied by the commutative subsequence. In the example of 4.4, for instance, ϕ_z (point 12) and ϕ_v (point 13) are unordered among themselves, but the set of them is linearly ordered in respect to ϕ_k (point 11) and ϕ_s (point 14).

While the trace of an operator is the same wherever the operator occurs

---

[34] In this way we can say that C operates on all S pairs, whereas otherwise we would have to say that it operates only on pairs satisfying certain conditions. And this extension is not arbitrary, because the cases of $S_1 C S_2$ are indeed derivable from $S_1 C S_2 CS \ldots CS$, by regular zeroing (5.6). C between arbitrary two S holds at least for the *and* and *or* of logic, which have to be included in the vocabulary of English.

in the partially ordered characterization of a sentence, the argument of the operator is different for each occurrence of it, and can be stated in terms of its position in the partial ordering.

The transformations impose two different and important partitions on the set of propositions (graded sentences).

One partition is according to the trace which each proposition contains: $A = B$ if A contains the same ordered traces as B. Whether ϕ_c are taken as unary or as binary, each sentence contains the traces of the partially ordered applications of particular base operators (and, possibly, analogic fixed sequences of these operators). Each graded sentence (proposition) is represented by one and only one sequence of ϕ. The relation R of having the same ϕ-product trace (without regard to the elementary sentences on which the ϕ operate) is an equivalence relation on the set P of propositions. The factor set P/R is the set of ϕ-products, since the traces are of ϕ applications. The mapping of the set P onto its factor set P/R, assigning to every proposition in P the ϕ-product in P/R whose trace that proposition contains, is the natural mapping of P onto P/R. The propositions which contain the trace of no ϕ are sent into the identity of the set of ϕ sequences, which we may write ϕ_k as above. These sentences are the kernel of the natural mapping, and will be referred to as sentences of the kernel, or kernel sentences, K. They are important because, for each graded sentence, they are the residual (elementary) sentences S_e under transformations.

The other partition is according to the residual sentences which are contained in each proposition under transformations: $A = B$ if A contains the same kernel-sentences as B. All sentences which are transforms (not only paraphrastic) of each other contain the same residual sentences. The empirical statement that sentences A_i, B_i are corresponding members (with same word choice) of sentence sets A, B which are transforms of each other is expressed in transformational theory by saying that $A_i \leftrightarrow B_i$ if and only if there exists a succession of base transformational operators which sends one of these into the other. As these operators have been defined, sentences differing only in operators must have the same residual sentences. These residual sentences of each sentence contain no transformational trace, and are therefore in the kernel of the natural mapping above. In the partition of the set of propositions into subsets whose members are transforms of each other, each subset contains one or more kernel sentences, which are contained in each sentence of the subset. These kernel sentences generate the subset by means of the transformational operators.

This transformational structure of sentences makes it possible to define sentences as those objects among which the ϕ relation obtains. A word sequence is a sentence if it has a ϕ relation to a sentence.

4.3.3. Prime sentences

We can try to reformulate this decomposition of each sentence into kernel sentences and operators so that it becomes a decomposition into prime sentences. To do this, we would have to replace each operator (elementary, or analogic fixed sequence) by an existing sentence of the language which would occupy the position of the operator in decompositions of sentences. Since the resultant of each operator is a sentence, we can form a "carrier" sentence for that operator out of that operator itself plus pro-words of its operand:[35] e.g., for ϕ_v we form *The subject V_v doing it* (so that in place of the operator *begin* we use the carrier sentence *The subject begins doing it*). If the carrier sentence thus formed contains nothing but this, it cannot appear in any sentence decomposition except at points where the corresponding operator would appear. The only problem becomes the availability of the required pro-words in the language, especially in the case of operators whose domain is a particular subclass of words (e.g., *take a ... n* in *He took a walk*, *He took a look*, etc.) for which no restricted pro-word can be found. In many languages the replacement of operators by carrier sentences can be carried out only by the introduction of highly artificial carriers. In the case of English, for example, we might have:

operator type	operator	carrier sentence for the particular operator
ϕ_v	... is -ing ...	He/she/it is doing it.[36]
	... begin to ...	He/she/it begins to do it.
	... take a -n ...	He/she/it takes the action.
ϕ_s	I know that S	I know that he/she/it does it (or did, will, etc.).
	Sn is slow	His/her/its doing of it is slow.
ϕ_c	and	This is also the case.
	because	The former is because of the latter.

[35] I.e., a resultant of that operator on a dummy sentence, such as *The subject does it*.
[36] The morphophonemic choice among these three is automatically determined by the subject of the operand sentence. The tense of ϕ_v sentences is the same as that of their operand.

wh-		carrier for *and* or *if*, plus: *N of location n-i in discourse is same individual as N of location n-j.* (Here *n* is the location of the *wh*-carrier in the discourse.)
ϕ_p, ϕ_z	—	instructions to carry out the particular operation
analogic fixed sequences	passive	*It is done by him/her/it.*

Such carrier sentences for English are clumsy, and some (especially for *wh-*, ϕ_p, ϕ_z) are metalinguistic. They lack the linguistic properties of kernel sentences (e.g., of containing only words of a particular type, concrete). However, they may have other properties of their own, such as containing only pro-words and a ϕ trace in the case of one type. Another type gives a fixed relation between stated words of neighboring primes, e.g., the *wh*-carrier (5.7.2).

We can describe each proposition of English (including the carrier sentences themselves) as uniquely decomposable, i.e., factorizable, into such carrier sentences and kernel sentences, partially ordered. These carrier and kernel sentences therefore constitute the primes of the set of sentences. E.g., *A young boy's beginning to walk is slow* =

A, *A boy walks.*
B, *A boy is young.*
C, *This is also the case.*
D, *The word in n-2, 2 represents the same individual as the (same) word in n-3, 2.* (Here D = n = 4.)
E, *'wh-is' is zeroed* (with automatic permutation of remainder).
F, *He/she/it begins to do it.*
G, *His/her/its doing of it is slow.*
CBA yields *A boy walks and a boy is young.*
DCBA yields *A boy who is young walks.*
EDCBA yields *A young boy walks.*
FEDCBA yields *A young boy begins to walk.*
GFEDCBA yields the given sentence.

The decomposition is partially ordered. B comes after A; for the sequence of primes *A boy is young. A boy walks*, followed by C, D would yield *The boy who walks is young.* G comes after F; for if it came before we would obtain *A young boy's walking is slow* (and with later F: *A young boy's walking begins to be slow*). Each of C, D, E, and the ordered set A, B, act in some one order, but the trace which they effect could be defined as due to any order of them, since the ordering of C, D, E is determined by the metalinguistic references they contain. And the ordered set F, G is unordered with respect to all these.

One can investigate the properties of the decomposition into primes, for all sentences, or for those in distinguished subsets. It is of interest to see how the differences between decomposition into primes here and those in the set of natural numbers relate to the great differences between the set of sentences and the set of numbers, or to the differences between the algebraic structures that can be usefully defined on each of these sets. In the set of sentences, the number of primes is finite; neither the primes nor the sentences as a whole are ordered, in any relevant way that has been noted so far. Furthermore, the decompositions of sentences are partially ordered, and the requirement of matching the resultant and the argument of successive primes (operators) in a decomposition means that certain combinations of primes do not occur in any decomposition, i.e., do not make a sentence. Thus we have decompositions containing kernel primes and carrier primes (as in the example above), and carrier primes alone (as in *I know that the latter is only because of the former*)[37] and one kernel prime alone (as in *A boy walked*). But no sentence contains more than one kernel prime without also containing a carrier prime of the ϕ_c type for each kernel prime after the first. These restrictions on the combinability of primes may be varied in interesting ways, or eliminated, for suitable subsets of the set of sentences (e.g., all sentences containing only one kernel sentence), or for certain altered definitions of the primes. For an example of the latter, the carrier primes for ϕ_c could be based on a unary rather than binary treatment of ϕ_c. Then, instead of B, C, D, above we would have B*, *The same is young*, as carrier for *wh-* plus any sentence of the form *N is young*.

Various subsets of sentences can be defined in respect to decomposition. For example, we can consider the set of sentences modulo their first kernel prime: i.e., all sentences whose first prime is, say, *A boy walked*; then all whose first prime is *A man walked*; etc. Any two such subsets whose first kernel prime is of the same kernel type (e.g., *NtV, NtVN* of particular

---

[37] Since the carrier sentences are taken as indecomposable primes, the set of carrier sentences differs from the set of operators (elementary and fixed sequences). E.g., ϕ_z acts only in sentences which have ϕ_c or ϕ_s. But a carrier sentence of ϕ_z, e.g., E in the above decomposition, can appear as an English sentence, indecomposable, and requiring no ϕ_c.

word subclasses, etc.) will contain the same decompositions; there will be an isomorphism of the first subset onto the second preserving carrier products.

4.4. *Decomposition lattices*

We have seen that each sentence has a partially ordered decomposition into elements, which may be taken either as prime sentences or else as base operators and kernel sentences. The decomposition is unique for each proposition, if the analogic transformations are taken as single elements. There are certain restrictions on the combinations of primes, or of ϕ, that occur in a decomposition; certain combinations occur in no decomposition.

With the elements taken as kernel sentences, unary and binary base operators, and analogic transformations, each proposition of the language can be written uniquely as a sequence of element symbols requiring no parentheses, for example in the manner of Polish notation in logic. Certain subsequences are commutative (representing elements unordered in respect to each other).

For sentences which contain only one kernel sentence, the decomposition can be represented as a nonmodular, nondistributive lattice with the kernel sentence as null element and the given sentence as universal element. The points represent the operators: c is the least upper bound of a, b, if c is the first operator which can be applied after both a and b (on the resultant of a, b) on the way from the kernel sentence to the given sentence, in this decomposition; and correspondingly for the g.l.b. The operator at the universal element can be taken as the ϕ_m which introduces the final sentence intonation; or it is f^4 if the lattice is made with the entities of 7.1.2. The same representation holds also for all other sentences, i.e., for those containing more than one kernel sentence, if the ϕ_c are taken as unary operators, $\phi_c: S_1 \to S_1 C_k S_{k1}$.[38] If the ϕ_c are taken as binary operators, $\phi_c: S_1, S_2 \to S_1 C S_2$, then the decomposition of a sentence containing more than one kernel is a semi-lattice; but then each C is oriented, with a distinction of right-hand and left-hand, because $S_1 C S_2$ (which would be the union of ordered S_1, S_2) is a sentence related, but not generally equivalent, to $S_2 C S_1$ (which would be the union of ordered S_2, S_1). In a given semi-lattice, we take the l.u.b. of ϕ_m (which produces S_1) and ϕ_n (which produces S_2) as being C_{i12} (i.e., C_i producing $S_1 C_i S_2$) or as being C_{i21} (i.e., C_i producing $S_2 C_i S_1$). But if we want to compare various

---

[38] In this case, each unary ϕ_c point on a lattice, which adjoins $C_k S_{k1}$ to S_1, is itself the universal element of another lattice, namely the one which decomposes S_{k1}.

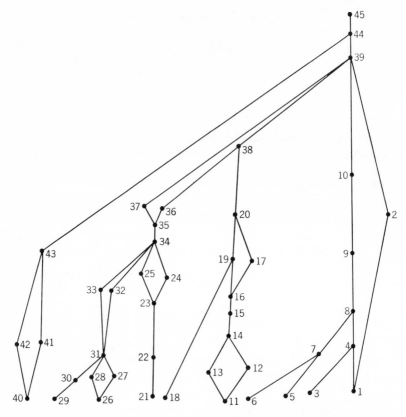

Figure 4.1

1. ϕ_k: *Adrenal is a gland.*
2. $\phi_m \phi_c \phi_k$: *the.*
3. ϕ_k: *A gland is endocrine.*
4. ϕ_c: *wh-.*
5. ϕ_k: *A gland is prime.*
6. ϕ_k: *A gland is endocrine.*
7. ϕ_c: *wh-.*
8. ϕ_c: *wh-.*
9. ϕ_v: *appears as.*
10. $\phi_m \phi_s$: *more and more.*
11. ϕ_k: *Adrenal is important.*
12. ϕ_z: Pronoun: *adrenal → it.*
13. ϕ_v: → *Adrenal has importance.*
14. ϕ_s: → *Adrenal's importance grows.*
15. ϕ_v: → *Adrenal's important has growth.*
16. ϕ_s: → *Growth of adrenal's importance is along a scale.*
17. ϕ_m: → *Adrenal's importance grows along* (or: *in*) *a scale.*
18. ϕ_k: *Animals are (ranged) in a scale.*
19. ϕ_c: *wh-.*
20. $\phi_m \phi_c \phi_k$: *the.*
21. ϕ_k: *A mammal lives.*
22. ϕ_v: → *A mammal has life.*
23. ϕ_s: → *Adrenal is indispensable for a mammal's life.*
24. ϕ_z: Pronoun: *mammal → it.*
25. ϕ_v: *has become.*
26. ϕ_k: N_i *removes adrenal.*
27. ϕ_v: → N_i *effects removal of adrenal.*
28. ϕ_z: Pronoun: *adrenal → it.*
29. ϕ_k: *A mammal dies.*
30. ϕ_v: → *A mammal suffers death.*
31. ϕ_c: → N_i*'s removal of it leads to a mammal's death.*
32. ϕ_z: on N_i*'s.*
33. $\phi_m \phi_s$: → N_i*'s removal of it leads rapidly to a mammal's death.*

semi-lattices in which C_i appears as l.u.b. of ϕ_m, ϕ_n, we have to add a property of orientation to the figures in order to show a different decomposition for the two different sentences $S_1 C_i S_2$ and $S_2 C_i S_1$.

The set of all lattices and semi-lattices, one for each proposition, has certain properties, i.e., certain dependencies among the occurrences of particular operators in particular relative positions within a lattice. These are different in part from the dependencies in a decomposition into primes. The most obvious one is that for each lattice, whose universal element is an arbitrary sentence, the null element is always a kernel-sentence. For each kernel-sentence beyond 1 there is precisely one ϕ_c point which is its l.u.b. with one of the other kernel-sentences in the semi-lattice.

As an example of the decomposition of a sentence (Fig. 4.1) we *take* the following:

The adrenal appears more and more as a prime endocrine gland: its importance grows in the animal scale: among the mammals, it has become indispensable for life, its removal leads rapidly to death: its functions are multiple.

A characteristic of transformational analysis is that languages are rather similar in their transformational structure, and that given a sentence in one language and its translation in another, the decomposition of each sentence in terms of the transformations of its language will be quite similar. An example is the analysis (Fig. 4.2) of the Korean translation of the first part (itself a whole sentential structure) of the sample sentence above.[39]

---

34. ϕ_c: $\rightarrow$ *It has become indispensable for a mammal's life, removal of it leads rapidly to a mammal's death.*
35. $\phi_m \phi_s$: $\rightarrow$ *... among mammals.*
36. $\phi_m \phi_c \phi_k$: *the* on *mammals.*
37. ϕ_z: on second *a mammal's.*
38. ϕ_c: semicolon.
39. ϕ_c: semicolon.
40. ϕ_k: *A gland functions.*
41. ϕ_z: Pronoun: *gland* $\rightarrow$ *it.*
42. ϕ_v: $\rightarrow$ *A gland has functions.*
43. ϕ_s: $\rightarrow$ *Its functions are multiple.*
44. ϕ_c: semicolon.
45. ϕ_m: sentence intonation.

[39] This analysis is the work of Maeng-Sung Lee.

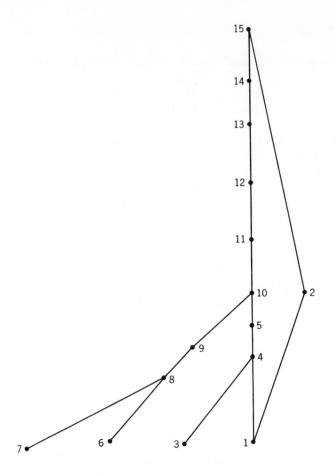

təuktə,	*pusin* ɯn	*cuŋyoha n*	*nǽpunpi sən* ɯlo	*poi ə ciko iss ta*
more and more	adrenal	principal	endocrine gland	seem/appear

More and more the adrenal is coming to appear to be a principal endocrine gland.

Figure 4.2

1. ϕ_k : *pusin-i sən-i-ta* 'Adrenal is a gland.'
2. ϕ_a : ɯ n on *pusin* (marks the topic-word of the sentence).
3. ϕ_k : *sən-i nǽpunpi-lɯl ha-n-ta* 'A gland does internal secreting.'
4. ϕ_c : E-nominalization (equivalent to *wh-*): $\rightarrow$ *pusini nǽpunpilɯl hanɯn sənita* 'Adrenal is a gland which does internal secreting.'
5. ϕ_z : on the appropriate verb *hata*: $\rightarrow$ *pusini nǽpunpisənita* 'Adrenal is an endocrine gland.'
6. ϕ_k : *sən-i cuyoha-ta* 'A gland is principal.'

7. ϕ_k : same as 3.

8. ϕ_c : E-nominalization: → *nærpunpilul hanun səni cuyohata* 'The gland which does internal secreting is principal.'

9. ϕ_z : on *hata*: → *nærpunpisəni cuyohata* 'The endocrine gland is principal.'

10. ϕ_c : E-nominalization: → *pusini cuyohan nærpunpisənita* 'Adrenal is a principal endocrine gland.'

11. ϕ_v : *poi-ta* 'appears': → *pusini cuyohan nærpunpisənulo pointa* 'Adrenal appears to be a principal endocrine gland.'

12. ϕ_v : *ci-ta* 'comes to: → *pusini cuyohan nærpunpisənulo poiəcinta* 'Adrenal comes to appear to be a principal endocrine gland.'

13. ϕ_v : *iss-ta* 'is ... ing': → *pusini cuyohan nærpunpisənulo poiəciko issta* 'Adrenal is coming to appear to be a principal endocrine gland.'

14. ϕ_a : adverb *təuktə* 'more and more.'

15. ϕ_m : sentence intonation.

5

Structures defined by transformations

5.0.

The establishment of the transformational method makes possible certain further analyses of language. In the first place, the transformational relations among sentences can now be used to investigate what relations there are among successive sentences. For sentences connected by a conjunction, we find a regular source which throws much light on the meaning of connectives (5.6). When we study how sentences refer to items elsewhere in the text, or to items apparently not in the text, we find that reference can be characterized as a particular relation among sentences of a discourse (5.7). Finally, one can study how the successive sentences of a discourse differ in their K and incremental ϕ composition; this leads to characterizing discourse as involving a second dimension beyond linguistic linearity (5.8).

These extensions of the analysis may not have the obviousness of the purely sentential analysis, but they provide regular and explanatory sources for the more complex structures in language.

Furthermore, the fact that all transformations are obtained as products of a few families of base transformations, each of which has a simple form and a reasonable semantic effect, makes it possible to modify the form of language with predictable semantic effect.

Here we will sketch certain modifications which detract little from the power of language. In particular: a set of unambiguous sentences or sentence pairs which is homomorphic to the set of sentences (5.1), a set of sentences containing no paraphrases which is isomorphic to the set of paraphrase subsets (5.2), and the set of those sentences which are obtained only from the monoid of free products of the base ψ transformations and products including ϕ-ψ, and which lack only certain paraphrases and extendabilities from the whole language (5.3).

At the same time, the transformational system makes it possible to understand various special subsets of language, and to carry out and evaluate various organizations of the set of sentences. Thus we can construct, from the transformational system of the language, the grammar of its metalanguage and obtain some of the properties of the metalanguage (5.4); one can obtain science sublanguages which have a grammar intersecting that of the whole language and which may have a relation like that of ideals (under conjunctions) to the language (5.9), and a graph of the inclusion relations of all sentences which leads to the characterization of a finite sentence containing all the information stateable in a language at a given time (5.5). And one can construct grammars for pairs of languages, in a way that provides a framework for differences among languages (5.10).

5.1. Unambiguous subsets of sentences

If the intersection of word subclasses were empty, and if there were no degenerate results from the paraphrastic ϕ, there would be no grammatical ambiguity; the only ambiguity in language would be due to the spread of meanings of the words in the elementary entities (kernel sentences and base operators). However, in natural language, it is frequently the case that different classes or subclasses of words, distinguished by their being part of different elementary entities, have some members in common. This permits a vast reduction in the size of vocabulary needed for a language. However, it can result in degeneracies in the set of produced sentences, i.e., it can result in the same sequence of words being produced by two or more different partially ordered sets of operators and kernel sentences. Such degeneracies are a factor in the development of analogic transformations.

If we wish to form a language without grammatical ambiguity, we can associate each ambiguous sentence with its transformational decompositions (5.1.1), or with a distinguished partial sentence of it (5.1.2); in the latter case we have an unambiguous language which consists only of sentences (more precisely, sentence pairs) and not of analyses.

5.1.1. Unambiguous decompositions of ambiguous sentences

Given a word sequence which is a grammatically ambiguous sentence, the two or more propositions which are expressed by the word sequence differ in their decomposition semi-lattices. These semi-lattices differ at two or more linearly ordered points, the later-operating point obliterating the difference brought in at the earlier point. For if two lattices have different operators at only one point, and yet produce the same word sequence, it would have to be the case that the different operators at that point in the two lattices introduce the same words in the same positions of the sentences under construction; but these words would be appearing, in the two sentences, as members of different operators which act in the same way in the order of operators. Whether this is possible, i.e., whether a given language contains two identically operating transformations, ϕ_1 and ϕ_2, such that there exists a member of ϕ_1 consisting of the same trace as some member of ϕ_2, can be seen from the list of transformations for the language. In English, this does not seem to happen. If, then, we disregard this possibility, there will have to be two or more differences between two decomposition semi-lattices of an ambiguous word sequence.

Any two decompositions of an ambiguous sentence must be similar to each other (except for the differences specified below); otherwise they would not yield the same word sequence. In most cases, they differ in a kernel sentence and also in one or more transformations (which obliterate the difference in kernel sentence). For example, in the ambiguous *Frost reads smoothly*, one decomposition is (omitting some details):

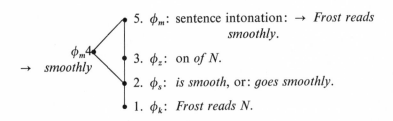

5. ϕ_m: sentence intonation: → *Frost reads smoothly*.

3. ϕ_z: *on of N*.

2. ϕ_s: *is smooth*, or: *goes smoothly*.

1. ϕ_k: *Frost reads N*.

ϕ_m4

→ *smoothly*

N is used here for a disjunction of all *N* which might occur in this position of the kernel sentence; the disjunction could be pronounced as *anything*, *things*, etc. (Properly, this results from ϕ_c: *or on many K*.)

After point 2 the resultant is:

Frost's reading of N is smooth (or: *goes smoothly*).

After 3, with zeroing of the disjunction *of N*,

Frost's reading is smooth (or: *goes smoothly*).

After 4, we have *Frost reads N smoothly*; but the combined effect of both 3 and 4 is:

Frost reads smoothly.

The other decomposition is:

- 5. ϕ_m: sentence intonation: $\rightarrow$ *Frost reads smoothly*
- 4. ϕ_m: $\rightarrow$ *smoothly.*
- 3. ϕ_z: on *N's*
- 2. ϕ_s: *is smooth* (or: *goes smoothly*).
- 1. ϕ_k: *N reads Frost.*

N is used here as above, but it covers a different subset of nouns, and the corresponding pronoun would be *anyone*. As above, ϕ_z is the zeroing of a disjunction, but it applies to a different disjunction in a different position.

After 2, the resultant is:

> *N's reading of Frost is smooth* (or: *goes smoothly*).

After 3, it is:

> *The reading of Frost is smooth* (or: *goes smoothly*).

And after 4, it is:

> *Frost reads smoothly.*

In some cases, the kernel sentences of both decompositions are the same, and all differences are in the transformations. This requires that the importation or change of words by some succession of operators in one decomposition be the same as that due to some other succession of operators in the other decomposition. This is possible, since there are several cases in which a particular word or morpheme (e.g., *be*, *-ing*) appears in more than one transformation. Indeed, one of the motivations in defining the base operators was to have each word or morpheme occur in only one operator, so that any occurrence of the word or morpheme would indicate an application of that operator. This result could not be completely achieved in the case of English as long as we want the base operators to be themselves transformations (4.2.4). Hence it is possible that identical traces may be produced by different successions of ϕ, especially if some of the ϕ are ϕ_z which may zero parts of the trace that is due to preceding ϕ of the succession. Any such situations which can arise can be determined from an inspection of the ϕ list and ϕ product table for the given language. A particular case, and the only one important in English, is when one decomposition contains $\phi_{z(i)}\phi_j\phi_i$, where $\phi_{z(i)}$ is a ϕ_z which precisely zeroes (or pro-words) the trace of ϕ_i, the lost record of ϕ_i being then reconstructable from the presence of the ϕ_j trace, or from the fact that the resultant is ambiguous. Since ϕ_z occurs only in specifiable ϕ successions, and has specifiable effects (which may include the elimination of part of the trace of ϕ_i), the cases of $\phi_{z(i)}\phi_j\phi_i$ can be specified for a given language. In English, the main case is when the ambiguous word sequence results from a later ϕ_c on an earlier ϕ_c or ϕ_s. The ambiguity arises from the fact that either only one of the kernel sentences which are under the later ϕ_c had the earlier ϕ (ϕ_c or ϕ_s), or that both had the same earlier ϕ but the trace of one occurrence of it was zeroed or pro-worded by the ϕ_z which followed the later ϕ_c.

For a simple example, we consider *A man walked and talked slowly.* One decomposition is:

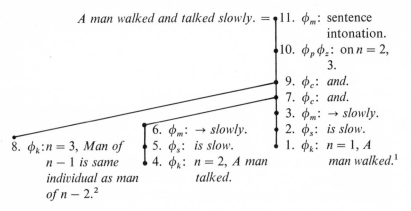

A man walked and talked slowly. =
- 11. ϕ_m: sentence intonation.
- 10. $\phi_p\phi_z$: on $n = 2$, 3.
- 9. ϕ_c: *and.*
- 7. ϕ_c: *and.*
- 3. ϕ_m: → *slowly.*
- 2. ϕ_s: *is slow.*
- 1. ϕ_k: $n = 1$, *A man walked.*[1]

6. ϕ_m: → *slowly.*
5. ϕ_s: *is slow.*
4. ϕ_k: $n = 2$, *A man talked.*

8. ϕ_k: $n = 3$, *Man of n − 1 is same individual as man of n − 2.*[2]

The resultant sentences after each point in the semi-lattice are:

After point 2: *1, A man's walking was slow.*
After point 3: *1, A man walked slowly.*
After point 5: *2, A man's talking was slow.*
After point 6: *2, A man talked slowly.*
After point 7: *1, A man walked slowly and 2, a man talked slowly.*
After point 9: *1, A man walked slowly and 2, a man talked slowly, and 3, man of 2 is same individual as man of 1.*
After point 10: where ϕ_z replaces K3 by the zeroing of those words in K2 which are identical with corresponding words in K1 (in this case *a man, slowly*) and the permuting of the residue of K2:

A man walked and talked slowly.

---

[1] For the numbering of the kernel sentences, see 5.6.
[2] For this form, see 5.6. *K* here stands for *K* with its ϕ_s.

The other decomposition is:

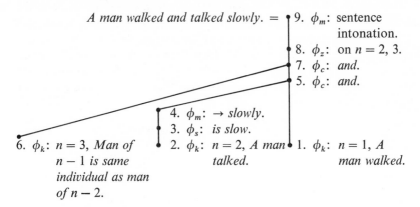

A man walked and talked slowly. = 9. ϕ_m: sentence
                                          intonation.
                                     8. ϕ_z: on $n = 2, 3$.
                                     7. ϕ_c: *and.*
                                     5. ϕ_c: *and.*

4. ϕ_m: → *slowly.*
3. ϕ_s: *is slow.*

6. ϕ_k: $n = 3$, *Man of* 2. ϕ_k: $n = 2$, *A man* 1. ϕ_k: $n = 1$, *A*
   n − 1 is same   *talked.*   *man walked.*
   individual as man
   of n − 2.

After point 3: *2, A man's talking was slow.*
After point 4: *2, A man talked slowly.*
After point 5: *1, A man walked and 2, a man talked slowly.*
After point 7: *1, A man walked and 2, a man talked slowly and 3, man*
               of 2 is same individual as man of 1.
After point 8: where the zeroable words are in this case only *a man*:
               A man walked and talked slowly.

The differences between the two semi-lattices are in point 2 (and 3) of the first one, which are lacking in the second, and in the difference in argument of ϕ_z as between point 10 of the first semi-lattice and point 8 of the second.

5.1.2. *The distinguishing partial sentence*

Instead of differentiating the two decompositions of an ambiguous sentence by the two or more points of difference in each, we can, alternatively, differentiate them by a distinguished partial sentence in each. In a semi-lattice presenting a decomposition of S_1, the resultant at point m (i.e., the sentence formed at point m by the kernel sentences and the operators at all points k, $k \leq m$) will be called a partial sentence of, or grammatically included in, the resultant of each semi-lattice point n, $m < n \leq S_1$.

Two semi-lattices of an ambiguous sentence have different partial sentences over a stateable section: from the earlier-operating point of difference to the later-operating point of difference. Each of the two semi-lattices, and hence each of the two propositions which they decompose, can be distinguished by the least (most included) partial sentence which does not appear in the other semi-lattice. If the semi-lattices differ in kernel sentences, the different kernel sentences are the distinguishing partials. If not, and one of the semi-lattices contains the $\phi_{z(i)}\phi_j\phi_i$ sequence, then it is distinguished by the ϕ_i resultant, while the other semi-lattice is distinguished by the first partial sentence in which the lack of the ϕ_i is recognizable (or, for simplicity, by the kernel sentence which fails to receive the ϕ_i). Hence for each n-fold ambiguous sentence we can form n subsets of sentences, each subset consisting of that sentence and the distinguishing partial sentences on its mth decomposition, for each m, $1 \leq m \leq n$. E.g., for the examples above, the sentence pairs are:

1. {*Frost reads smoothly; Frost reads N*},
2. {*Frost reads smoothly; N reads Frost*};

1. {*A man walked and talked slowly; A man walked slowly*},
2. {*A man walked and talked slowly; A man walked and a man talked slowly*}

or for simplicity:

2. {*A man walked and talked slowly; A man walked*}.

We now form a set of subsets of sentences as follows: For each ambiguous sentence, a subset is formed for each different decomposition of the sentence, consisting of the given sentence and the differentiating partial sentences of that decomposition, these being in some cases simply the kernel sentences and in other cases the kernel sentences with a particular ϕ. For all nonambiguous sentences, the subset is simply the given sentence itself. This set of subsets of sentences contains no ambiguities, and maps isomorphically onto the set of propositions of the language.[3] This set of subsets can be obtained from the set of sentences by means of the base operators, and provides us with a set of sentence pairs (possibly sentence subsets) containing no ambiguities, which is expressed purely in terms of sentences (subsets of sentences) without explicit reference to the analyses or gradings of the sentences.[4]

---

[3] The isomorphism with the set of propositions depends on the fact that no word sequence can appear twice in an acceptability-graded subset of sentences. If a sentence is n-way ambiguous, it appears in n different acceptability-graded subsets of sentences, i.e., in n partitions of the set of propositions.

[4] Since ambiguous sentences have various transformational decompositions corresponding to their various grammatical meanings, and since the traces of these various transformations affect the word repetition which is required in all regular $SCS \dots CS$ and discourse neighborhoods in which the ambiguous sentence appears (5.6), it follows that each n-way ambiguous sentence appears in n different sets of $SCS \dots CS$ or discourse neighborhoods. The neighborhood therefore differentiates the different meanings of an ambiguous sentence (i.e., a sentence is not ambiguous in its neighborhood), unless the differentiating parts of the neighborhood have been zeroed.

5.2. *Partition into paraphrastic subsets*[5]

The nonincremental operators ϕ_p, ϕ_z, ϕ_m have various effects which are convenient for the construction of complex sentences: especially ϕ_z which yields vast reduction in the length of sentences. However, these and the analogic transformations bring into the language sentences which are only paraphrases of other sentences. For some purposes it may be desirable to construct a language without paraphrases, i.e., one in which if two sentences differ in form they differ in meaning. This would be at the cost of the advantages which the paraphrastic transformations had brought. It is possible to construct a subset of the set of sentences in which there is no grammatical synonymity, i.e., in which all different sentence forms of a given word choice (*n*-tuple of word values) differ as to information.

We ask which transformations produce paraphrases, such that ϕS_1 carries the same information as S_1, aside from considerations of stress and style. Clearly, the ϕ which do not operate on K produce paraphrases:

ϕ_p (*Little would I expect him to go.* ← *I would little expect him to go.*);

ϕ_z (*People came and went.* ← *People came and people went*) though ambiguity is introduced if $\phi_z S_2$ is the same word sequence as $\phi_z S_1$;

ϕ_m (*I asked whether he came.* ← *I asked: Did he come?*)

In addition, certain informationally empty and hence zeroable ϕ_{s_0} (including the performatives) produce paraphrases: e.g., *They picketed the plant.* → *Their picketing of the plant took place.*

Also paraphrastic are the analogic fixed sequences which consist of ϕ_{s_0}, particular informationally empty ϕ_{c_0} (*wh-*, *and*), and ϕ_p, ϕ_z, and inverses of these: *He purchased books; His purchase was of books; The purchase which was by him was of books;* etc.

All these are paraphrastic because the traces which they introduce do not add to the information[6] (and, in view of the recoverability of zeroed material, do not subtract from it).[7] The ϕ_p and ϕ_z give some indication of the relation of the given sentence to its neighboring sentences, but do not

[5] See also H. Hiż, The role of paraphrase in grammar, *op. cit.* in Chapter 1.

[6] As to the other words of the paraphrastic sentences, these are of course the same, since the sentences are transforms of each other.

[7] Correspondingly to footnote 4, it may also be expected that the neighborhoods of paraphrastic sentences are identical or similar in a way that the neighborhoods of nonparaphrastic sentences are not (see Chapter 8, footnote 3), so that neighborhood in discourse can be used as criterion of paraphrase, instead of using judgment as to whether difference of meaning exists. However, this has not been investigated.

change the information in the given sentence. The only nonparaphrastic transformations therefore are the incremental ones (except when they occur in analogic sequences, as above): ϕ_a, ϕ_v and ϕ_s as in *He just slept, He tried to sleep, His sleeping was quiet*; ϕ_c as in *He slept after they phoned* (all from *He slept*).

While the discussion here is of paraphrastic sentence forms, we must note that many actual propositions are paraphrases of each other, not on the grammatical basis of the above transformations but on the vocabulary basis of local synonymity of words: *He spoke. He talked.* In this case the paraphrases have different word choices, rather than (as above) the same ones. Every sentence has such synonymity paraphrases, because every word or idiomatic word sequence in a language has local synonyms (including dictionary definitions) in each of its neighborhoods. This word synonymity may be eliminable by the method of 6.5.

Sentences A, B are paraphrases of each other if A, B differ only by paraphrastic sequences of transformations (or by local synonyms). If A, B are paraphrases of each other except that where A has N_i, B has a pronoun or classifier $N_{cl(i)}$ of N_i, then the difference between A and B is only the pronoun operator or the added prime sentence N_i is a (*case of*) $N_{cl(i)}$. It is therefore possible to inspect the decomposition lattices of two sentences and to say whether they are grammatical paraphrases of each other or not.

The relation of being a paraphrase, whether due to transformations or to local synonymity, is an equivalence relation in the set of propositions. It yields a partition of the set of propositions into paraphrase subsets such that within a subset all sentences are paraphrases of each other. All the sentences within a subset have in common, as against all of the other subsets, a particular ordered set of kernel sentences and of ϕ_a, ϕ_v, ϕ_s, ϕ_c (except ϕ_{so}, ϕ_{co}).

Each such subset contains at least one word sequence (proposition) which is not a member of any other of these subsets. This can be seen as follows: If a word sequence A is ambiguous, and is therefore a member of two or more of these paraphrase subsets, it has a different decomposition in each subset. But each decomposition of A has one or more partial sentences B which are unique to it and are not obtainable in the other decompositions of A. The ambiguousness of A results from certain zeroings or permutations E on B. If we construct from B a sentence containing all the further ordered operations which produced A, except for these particular zeroings or permutations E (which are themselves paraphrastic), we will obtain a sentence which is a paraphrase of A (being identical with A except for the paraphrastic operators E), but is not ambiguous. We can therefore select from each paraphrase subset, on the basis of some overall principle, one unambiguous sentence which can be taken as representative of the subset, and thus obtain a set of unique representative sentences which is isomorphic (under the incremental transformations) to the set of paraphrase subsets, and which therefore carries the same information as is carried in the whole set of sentences of the language. A language informationally equivalent to the natural one can be obtained if each incremental ϕ operates only on the representative sentence of each paraphrase set, the remaining members of the paraphrase sets being discarded from the language.[8]

Somewhat differently, we can take the kernel sentences of the language, and permit all possible applications of the nonparaphrastic transformations (i.e., we eliminate from the set of transformations all the paraphrastic transformations from among the base operators or the analogic fixed sequences). We thus obtain a set of sentences, none of which are grammatical paraphrases of any other, and such that all other sentences, outside this set, are paraphrases of one or another of the sentences in this set.[9]

Sentences which are paraphrases of one another may be substituted for one another in a discourse, without changing neighboring sentences in the discourse, except for possibly requiring paraphrastic transformations on these neighbors, for considerations of style and of convenience in stringing the sentences together in the discourse.

The elimination of the paraphrastic base operators ϕ_p, ϕ_z, ϕ_m leads to certain changes in the grammar of the remaining sentences, which are constructed out of the incremental operators ϕ_a, ϕ_v, ϕ_s, ϕ_c. For example, most ambiguous sentences (based on ϕ_z) disappear, and so do the analogic operations. Also, the formulation of the *wh*-connective becomes more complicated. The *wh*-connective requires that the second sentence begin with N_i (or PN_i), where N_i is an N that occurs in the first sentence: *wh:*

---

[8] This does not apply to metalinguistic sentences. The metalinguistic predicate *is N_{II}* (5.4) operating on S_1 does not give the same information as when it operates on a paraphrase of S_1. The relation of being a paraphrase does not hold in the metalinguistic form *X is N_{II}*, and under quotation marks which are derived therefrom. *He is here* and *They said that he is here* are paraphrases of *He is present* and *They said that he is present*, respectively; but *They said 'He is here'* is not a paraphrase of *They said 'He is present.'* Hence *is N_{II}* must be defined on all discourses and discourse fragments, and not merely on the representative sentences of the paraphrase sets.

[9] There are also sentences which are stated metalinguistically to be paraphrases of other sentences, e.g., in such sets of metalinguistic sentences as 'S_1' *is a paraphrase of* 'S_2' or 'S_1' *means* 'S_2' (*means* here not as synonym of *implies*). However, many or all of these S_1, S_2 pairs would be shown to be paraphrases by the methods described here and in 6.5, independently of such metalinguistic statements.

I saw the man, The man left → I saw the man who left. But in many cases N_i begins the second sentence only by virtue of ϕ_p: *wh: I found this book, This book I wanted → I found this book which I wanted.* In general, it would be necessary to state what properties of the transformations would assure that the representative sentences can accept the incremental operators precisely as all S could.

5.3. *Elimination of analogic transformations*

In the discussion of the analogic transformations (4.2.4), it was seen that the monoid of free products of ψ, and the $\phi - \psi$, did not contain all transformations. There remained certain extra transformations, each of which turned out on inspection to be stateable as a particular product of base transformations (or their inverses, which are not defined in the monoid), but one which violated at some point the condition that the counterdomain of the ith transformation in each free product intersect the domain of the $i + 1^{\text{th}}$ transformation. In all these extra transformations there was some i^{th} factor whose counterdomain was defined on some subclass A, while the $i + 1^{\text{th}}$ factor had a domain defined on some subclass A' similar but not identical to A. These extra products can be looked upon as extensions of the set of products of the base ϕ. However, only particular such extended (analogic) products exist in a language at any time. They spoil the regularity of operation of the base operators; and languages differ more in respect to the analogic transformations than in respect to the base operators. Furthermore, each of these analogic transformations is a paraphrase of some sentence which is derivable without it from regular products of base transformations.

In an unchanging language these analogic paraphrases are dispensable. It therefore becomes of interest to construct a set of sentences which is identical with the existing set except for the removal of all sentences which are due to the analogic transformations. This is obtainable simply by admitting no violations of the domain-inclusion condition; this will apply automatically to inverse operators based on that violation.

The set of sentences has then lost only certain paraphrases of certain of its sentences. The relations among the remaining sentences, and the formulation of their structure, are not affected (as they were affected in some cases in 5.2, e.g., in the definition of *wh*-). The only loss to the grammar is the loss of certain stylistic variations (such as the passive), and the loss of any possibility of extending the domains of the operators. The gain is that for this slightly reduced set of sentences, all sentences are produced out of K only by the monoid of products of ψ and by $\phi - \psi$.

5.4.   The metalanguage

The fact that the metalanguage is included in the language has major effects upon the properties of natural language. It means first of all that the language contains sentences about sentences. It affects the global picture of the membership of the set of sentences (5.5). Taken together with the power of ϕ_c, it means that metalinguistic sentences can be conjoined to nonmetalinguistic sentences. In particular, sentences which state all the distinguishing nonlinguistic context of a sentence A can be thus conjoined to A itself (5.6). And in conjunction with the linearity of sentences, it makes reference possible (5.7).

First we consider how, peculiarly for natural language in contrast to other systems, the metalinguistic sentences, i.e., the sentences which talk about sentences and sentence segments of natural language, are themselves sentences of the same natural language: e.g., *'He went home' is a sentence* is a sentence of the language. To show that the metalanguage is in the same language, we note first that, in English, all metalinguistic sentences contains transforms of the sentence form *X is a sentence, X is a word, X is a linguistic form of English*, etc., also *'X' is a sentence*, etc.

These metalinguistic sentences are thus seen to contain a subclass (written *is N_n*) of an *is N* predicate. This subclass contains *is a word, is a name*, etc., which name or characterize segments of the object language. Phonetically *'X'* indicates that X is pronounced with sentential or other special intonation, but we have to recognize that the object-language segments may be pronounced with or without quotes, as a matter of morphophonemic variation.[10] Hence we have:

He went is a sentence.

(by ϕ_m) → *'He went' is a sentence.*

and

Mary is a word.

(by ϕ_m) → *'Mary' is a word.*

[10] "Mention" applies to material within quotes, but it can be defined transformationally as originating precisely and only for X (without quotes) in the sentences *X is a name, X is a term*, etc., and in sentences which contain these (even if partially zeroed), such as *We call Y (by the name) 'X' ← wh-: We call Y by a name. The name is X.*, and *He is of the 'intelligentsia,' He is of what is called the intelligentsia ← wh-: He is of X. X is called the intelligentsia.* (cf. Chapter 4, footnote 24). In a given sentence, X is "mentioned" if and only if the decomposition of that sentence contains a kernel sentence *X is a name, X is a term, X is a word*, etc.

The ϕ_m here is a free morphophomemic operation which adds an intonation, written as quotes, before *is* $N_{,,}$, as it does after certain ϕ_s (*I ask*, etc.). Since this ϕ_m occurs specifically with *is* $N_{,,}$, the latter can be (recoverably) zeroed after ϕ_m has acted, somewhat as *I ask* (a member of ϕ_s) can be zeroed after the ϕ_m of question intonation has acted:

He will go. $\overset{\phi_s}{\to}$ *I ask whether he will go.* $\overset{\phi_m}{\to}$ *I ask: "Will he go?"* $\overset{\phi_z}{\to}$ *Will he go?*

Thus sentences like *'Mary' has four letters* are derived

by:	from:
(ϕ_z on *the word*)	← *The word 'Mary' has four letters*
(ϕ_z on *which is*)	← *The word which is 'Mary' has four letters*
(ϕ_c: *wh-*)	← *The word has four letters. The word is 'Mary.'*

There are metalinguistic sentences which do not contain a quotable segment of the object language but only an $N_{,,}$ classifier of these segments: e.g., *The word has four letters, English sentences contain verbs.* If we wish, we could derive such sentences by ϕ_z from *X is* $N_{,,}$ sentences, e.g., *'S_1' or 'S_2' ... or 'S_n' which are sentences contain 'A' or 'B' ... or 'X' which are verbs.*

The subject of *is* $N_{,,}$ is not necessarily a noun of the language. It can be any word or linguistic expression, or any sound, as in *'Book' is a noun, 'Go' is a verb, '-ly' is a suffix, 'New books' is the subject, 'He went' is a sentence.*[11] In almost all other sentences of English the subject can only be a noun or a nominalized word or sentence.

We now note that the predicate *is* $N_{,,}$, which allows its subject the ϕ_m of quotes, exists in natural language even without the metalinguistic cases, as in *is a sound, is a noise.* In sentences of the form *X is a sound* or *'X' is a sound,* the *X* can be replaced by any sound (not only nouns of the language); and the resultant is a sentence which can enter into various transformations:

He heard 'X' ← *He heard the sound 'X'* ← *wh-: He heard the sound, The sound was X; 'X' and 'Y' are sounds;* etc.

---

[11] The subject of *is a sentence* is an unmodified (unnominalized) sentence, different from the subject in *That he went is a fact* or from the object in *I know that he went.* In *I know he went* we do not have an unmodified sentence as object, but ϕ_z from *that he went.* Note that *I said he went* is ambiguous: by ϕ_z from *I said that he went,* in which case the words actually said may have been different; and by different ϕ_z from *I said the words* (or *sentence*) *he went.*

It follows that all we need in order to construct (or characterize the syntax of) metalinguistic sentences is to include in the *is* N_n subclass of predicates certain names of sentences, sentence sequences, and sentential segments (*word*, etc.). We can call the set of these classifiers (names of language entities) N_{meta}, a subclass of N_n.

The metalinguistic sentences are therefore part of their own object language not only in that speakers of the language recognize them as such (i.e., '*He went*' *is a sentence* is empirically itself a sentence), but also in that certain of the word classes (which include N_n) and sentence structures (including X *is* N_n) and transformations (including quotes) of the object language suffice to characterize the metalinguistic sentences. The syntax of the latter contains nothing which is not found in the syntax of the object language, except insofar as the set of metalinguistic sentences is restricted to containing only the new subclass N_{meta} of N_n.

The metalinguistic sentences, being only those that contain these new members X *is* N_{meta} of X *is* N_n (even if obliterated by later zeroing), are a subset of the sentences of the language, although they contain, in the position of X, all sentence sequences and sentential segments.[12]

In later sections we will see that there are various grades of complexity in metalinguistic sentences. First, there are sentences of the form 'X' *is* N_{meta} with linguistic material in the position of X: '*Mary*' *is a word.; In certain environments, '*buy*' *is a synonym of* '*accept*.' These may be called metatype sentences, and it is impossible to talk of language material without citing it in such a sentence.[13]

Second, there are sentences of the form a, 'q' *in* 'X' *is* N_{meta} (5.6.3), where a is the ordinal number of q (among segments of the class of q) in a linguistic form X which contains q: *The word* '*book*' *in word-position 2 of* '*the book*' *is a noun, but* '*book*' *in word-position 3 of* '*They will book him*' *is a verb*. These may be called metatoken sentences; it is impossible to talk about the occurrence of linguistic material q within some larger linguistic material X without citing X and giving the position of q within X.[14]

Third, there are sentences of the form

$$n - 1, S_1 \text{ and } n, S_{meta(n-1, S_1)}$$

where in a discourse (with necessarily numbered sentences, see 5.7) there

[12] Material in the position of X in 'X' *is* N_n or transformed from there, will be said to be "cited."

[13] In scientific discourse, there occur metascience operators and sentences which are not metalinguistic (5.8).

[14] The distinction between type and token can be derived from meta-type and meta-token sentence forms.

is a metatoken sentence which cites the preceding parts of the discourses. It is impossible to refer to some A in a discourse without citing the discourse and the position of A in it. If the referring sentence is part of the discourse, then it cites the completed portion of the discourse of which it is part.[15] A metatoken sentence about S_1 which is adjoined to S_1, as in the example above, will be called a metasentence of S_1.

Each sentence S_i of the language has finitely many metasentences, i.e., metalinguistic sentences about it, $S_{meta(i)}$, which say that it is a sentence (of a particular kind) or that its segments are sentence segments (of various kinds).[16] Thus, each way of reading the decomposition semi-lattice of a sentence S_i is a sentence about S_i, different from any metalinguistic sentence about any other sentence. Each metalinguistic sentence in turn has finitely many sentences about it (which say it is in the language, etc.), and so on without bound. In particular, the sentence $S_{lat(i)}$ which reads the semi-lattice of S_i has a decomposition semi-lattice, which can be read by $S_{lat^2(i)}$, which is a metalinguistic sentence on $S_{lat(i)}$ and only secondarily on S_i; and so on. Thus for each sentence of the language there are in the language denumerably many metalinguistic sentences made out of it, which we can recursively construct.

5.5. Graphs of the set of sentences

The existence of a unique decomposition for graded sentences makes it possible to order the whole set of graded sentences in respect to their decompositions, and to investigate various relations among sentences and certain properties of the whole set of sentences.

---

[15] As will be seen in 5.7, a sentence cannot refer to itself. Hence if the nth sentence of a discourse speaks about the $n-1^{th}$, it does not refer to it as "the sentence before the present one" (since that is undefined). Rather it in principle cites the whole discourse $S_1 \ldots S_{n-1}$ and says that in such a discourse S_{n-1} has a certain property. That this property applies to the sentence preceding is a matter of our interpretation, when we see that the S_n in question indeed follows such a discourse $S_1 \ldots S_{n-1}$. The situation differs from pseudo-citing material which is to come in later sections of a discourse, e.g., *Hereafter we shall use the term X*. In this latter case, explicit reference must be made to an ongoing discourse (see fn. 33).

[16] The set of metalinguistic sentences of S_i, $\{S_{meta(i)}\}$, contains all the grammatical information about S_i, both meta-type and metatoken, including its decompositions (in terms of transformational and other methods of analysis), and also its local synonymities. $\{S_{meta(i)}\}$ is finite because S_i has finitely many parts. But the metalinguistic sentences relating S_i to other sentences, including statements of transformational difference, may be denumerably many.

5.5.1. Graph of partial sentences

There are various types of graph which represent the relations among sentences in such a way that properties of transformations or of subsets of sentences can be derived from properties of the graph. Here we describe a graph which gives the grammatical inclusion relation among all propositions (i.e., graded sentences) of the language.

Consider the decomposition semi-lattice of an arbitrary proposition S_1. Here the vertices are operators (including the null points which select the kernel sentences, and the universal point which adds the sentence intonation); the edges connecting the vertices represent a minimal set of independent partial sentences of S_1 sufficient to characterize S_1. Because of the partial ordering of the transformations, some of the partial sentences of S_1 do not appear in the semi-lattice. We construct a graph representing all the partial sentences of S_1, i.e., all the sentences which can be obtained in the course of any order of decomposition of S_1, as follows: each operator vertex of the semi-lattice is replaced by an element consisting of a directed edge (representing the operator) from the end point of the preceding element, and a terminal vertex (which represents the resultant partial sentence); except that the null point is replaced by a vertex alone, and the universal point is replaced by nothing; in addition, each set of n lattice points which are unordered among themselves is replaced by $2^n - 2$ vertices representing all the sentences that can be formed by combinations of the n operators from 1 up to $n - 1$ at a time (we draw connecting edges first for each pair, then each triple, and so on, up to n-tuple of the n points; each set of connecting edges ends in a vertex representing the partial sentence).

In this directed graph of S_1, whose edges are transformations, all the vertices of the graph are all the partial sentences of S_1. The graph for each partial sentence S_q of S_1 gives all the partial sentences of S_q, and is a subgraph of the S_1 graph. A particular partial sentence S_p of S_1 may also be a partial sentence of some other sentence S_2. In that case the graph of S_p is a subgraph of the S_2 graph as well.

5.5.2. Longest sentence and finite information

We now consider the fact that every sentence is a partial sentence of certain other sentences. It is therefore possible to construct a single such graph for the set of all the propositions of the language. Each proposition is a vertex, and appears only once. It is the terminal vertex for all the transformations (edges) that produce it from its partial sentences, in all the ways in which this can be done, and it is the initial vertex for all the transformations which produce another sentence from it. Because the vocabulary and kernel forms and transformations are each only finitely many, there is a finite number of edge sequences $e_1, \ldots e_n$, such that for arbitrary e_i, $1 < i < n$, e_i reaches a particular vertex A_i after which all that can happen are repetitions of portions of the edge sequence entering A_i, starting with various preceding vertices; these can obviously be enumerated. We can then connect (by the *and* of logic) all the vertices A_i, $i = 1, \ldots n$, into a vertex U; the *and* of logic occurs between arbitrary S, and is available in natural language (5.6.1). The sentence represented by U is the informationally maximal sentence of the language, because it has the following properties: All further sentences which can be formed transformationally from it contain repetitions, to any number, of one or more distinguished parts of U; or an unbounded regress of metalinguistic sentences about metalinguistic sentences about U, its parts, and its extensions. Since the information contained in any repetition of a sentence part, or a metalinguistic statement about it, is stateable in terms of that sentence part, we can state the informational content of any further sentence in terms of the parts of U. Since U is finite, the number of its parts is finite. Except for these forseeable and enumerable further sentences built out of U, each proposition of the language appears once in the graph of U as a distinguished partial sentence of U and its graph is a subgraph of the U graph. The construction for U does not hold for those sentences in mathematical and other symbolisms which do not have a structurally equivalent translation in natural language.

Put differently: For a particular language at a particular time, it is possible to construct a graph of the longest (finite) sentence which does not contain iterations of distinguished parts of itself. It is then possible to describe in a (finite) sentence what are these distinguished iterable parts and what is the informational effect of iterating each of them (to any number of iterations).[17]

We thus have a directed graph of all the sentences of the language, and for informational interpretation a finite graph. Any sequence of edges connecting two sentences is a transformational path which takes us from one sentence to the other, where a transformation is taken positively when going in the direction of the edge, and inversely when going against the direction of the edge.

---

[17] E.g., one can state finitely the meaning of any number of iterations of *very* at any point in which it occurs; one can state in a (finite) sentence the meaning of n iterations of *son's* or of *son's daughter's* as a function of n. The sentences which describe the information of each iterable part can than be conjoined by logical *and* to make one sentence giving the information that is contained in all iterations of all parts of U.

5.6. Conjunctional sequences

5.6.0.

The possibility of restrictions beyond those of a sentence arises from the fact that there exist binary operations which produce one sentence, $S_1 C S_2$, out of two, S_1, S_2. Each of the component sentences satisfies the restrictions imposed by sentencehood. The question is whether the combined sentence $S_1 C S_2$ must satisfy certain restrictions beyond those satisfied by S_1, S_2 separately. That is, whether the subset of S found in $S_1 C S_2$ ... $C S_n$ (for $n \geq 2$) satisfies restrictions beyond those that apply to the whole set of sentences. It will be seen that it does, i.e., that C is not a simple binary composition in the set S, and that these additional restrictions are of considerable syntactic and semantic interest and usefulness.

The fact that each discourse can be rewritten as one long sentence, if in no other way than by inserting the *and* of logic between the successive sentences (or, if necessary, between paraphrastic transforms of them) might make one think that the restrictions due to C suffice to account for the restrictions due to discourse. However, this *and* introduces few if any restrictions, and it will be seen (5.8) that the restrictions due to discourse are further ones. That is to say, the set of discourses is a proper subset of the set of $S_1 C S_2 \ldots C S_n$ (when C is the *and* of logic which is replaceable by the period between successive sentences).

We have seen that discourses, i.e., the occurrences of material in a language, can be segmented into sentences (3.6), and that there are stated restrictions on the word sequences which constitute a sentence. In 5.8 we will see that a discourse contains certain further restrictions on word sequences, beyond those that are imposed by its component sentences. However, before investigating this we have to consider certain restrictions on word sequence which come after those due to sentencehood and before those due to the discourse. It will be seen that these restrictions are intermediate also in character between the other two.

5.6.1. Mimimal difference for C_o

We first note that certain conjunctions occur between two sentences only if there is at least a particular minimal difference between these. Thus, while *and* may be found between two identical sentences, *or* and *but* (or *whereas*) are not:

∃ *He will write it and he will write it.*

(even though the meaning is not "wrote twice" but "continued to write," which is relevant to the further analysis). But (except in logic):

> ∄   *He will write it or he will write it.*
> ∄   *He will write it but he will write it.*

We do not find *or*, in natural language, unless there is at least one difference:

> *He will write it or she will write it.*
> *He will write it or he will type it.*

nor *but*, *whereas* unless there are at least two differences (or, for *but*, certain contrasts in the predicate), even if one of the two differences is no more than the word *too* which contributes no substantive difference in meaning:

> ∄   *He will write it but she will write it.*
> ∃   *He will write it but she will write it too.*
> ∃   *He will write it but she will not write it.*

The requirements of word repetition (5.6.2) also apply to C_o, but not to the *and*, *or* of logic. (The latter are words of English, because the sentences of logic are included in English.)

5.6.2.   *Word repetition for C_s*

We next check to see if there are any restrictions due to the other C. To this end we define min acc (S_1, S_2) as the lesser of the two acceptabilities, of S_1 and of S_2, the acceptability of a sentence being determined from its position in the set of acceptance inequalities in which it participates. In the set of SCS, we can define the acceptabilities within each subset $S_1 C S_2$, for fixed S_1, in terms of the acceptability inequalities as we vary S_2. If for given S_1, S_2, S_3 and C_a:

$$\text{acc } (S_1 C_a S_2) = \text{min acc } (S_1, S_2)$$
$$\text{acc } (S_1 C_a S_3) < \text{min acc } (S_1, S_3)$$

we will judge that C_a imposes some restriction on the choice of S, such that the pair S_1, S_2 does not violate the restriction, but the pair S_1, S_3 does. First, we consider sentence pairs which have reduced acceptability, i.e., where

$$\text{acc } (S_1 C S_2) < \text{min acc } (S_1, S_2):$$

(1)   *He wrote poetry because it was Tuesday.*
                 The war will start unless he enters the room.

If we ask what, if anything, would raise the acceptability of such $S_1 C S_2$ so as to equal min acc (S_1, S_2), we would find, for example:

> *He wrote poetry because it was Tuesday, and he always writes poetry on Tuesdays.*
>
> (2) *He wrote poetry because it was Tuesday, and on Tuesday afternoon he has the poetry class.*
>
> *Since the invaders threaten war unless the Prince of Cambodia comes to their conference room, the war will start unless he enters the room.*

We find that the acceptability of these reduced acceptability $S_1 C S_2$ can be raised to equal min acc (S_1, S_2), and this by adding certain $CS \ldots CS$. Furthermore, for each such $S_1 C S_2$, many of the $CS \ldots CS$ which raise the acceptability have the property of repeating the main words of S_1, S_2. This suggests that the C imposes a restriction requiring word repetition.

True, there are cases in which the added $CS \ldots CS$ which raise the acceptability of $S_1 C S_2$ do not repeat the words of S_1, S_2:

> *He wrote poetry because it was Tuesday and that was his custom.*
> *He wrote poetry because it was Tuesday and he had to get his literary output over with before the middle of the week.*

Furthermore, we can find other $S_1 C S_2$ whose acceptability is not reduced in the first place, i.e., where acc $(S_1 C S_2)$ = min acc (S_1, S_2), even though they do not have word repetition:

> (3) *He wrote poetry because he was young.*
> *The opposition will go underground when the war starts.*

As a test, we try adding in all these cases the kind of $CS \ldots CS$ which were found to be useful in the first case, namely ones which contain the word repetition:

> *He wrote poetry because it was Tuesday and that, namely writing poetry on Tuesdays, was his custom.*
>
> *He wrote poetry because it was Tuesday and he had to get his literary output over with before the middle of the week, and writing poetry*
>
> (4) *is included in literary output, and Tuesday is before the middle of the week.*
>
> *He wrote poetry because he was young, and young people like to express themselves in poetry.*
>
> *The opposition will go underground when the war starts, because open opposition is impossible in wartime.*

We find that given $S_1 CS_2 \ldots CS_n$ which do not have reduced acceptability but also do not have word repetition, it is always possible to add $CS \ldots CS$ which, firstly, bring in the word repetition while maintaining the acceptability, and which, secondly, have the special property of being known statements of the dictionary or grammar, or of being common knowledge (that *that* in the first example is a pronoun for his writing poetry on Tuesdays, that Tuesday is before the middle of the week, etc.)

It remains to see why the word-repeating $CS \ldots CS$ additions raised the acceptability in the first case and merely maintained it in the second. The difference is that in (4) the $CS \ldots CS$ to be dropped give information which is already well known to all concerned, and so add nothing to the information in the $S_1 CS_2$ of (3), whereas in (2) the $CS \ldots CS$ to be dropped gave useful additional information. If we start from the longer (word-repeating) form, which is always acceptable, we can say that the conjoining of sentences has assured acceptability only if each main word occurs in at least two of the sentences. We can then derive the shorter form—in (3) but not in (1)—by saying that any CS which adds no information is zeroable, by an extension of the conditions stated in 4.2.2.6 for ϕ_z.[18] An $SCS \ldots CS$ which, like (3), has lost its zeroable CS is an extended ϕ_z transform of the original, and so remains unchanged as to acceptability, since transformations preserve acceptability ordering.[19] But an $SCS \ldots CS$ from which there have been dropped CS which do not satisfy the conditions for extended ϕ_z is no longer a transform of the original $SCS \ldots CS$ (but rather some informal abbreviation of it); this was the case in (1).

The words which remain unrepeated [e.g., in (4): *wrote*/ *express themselves; go underground*/ *open* ... *impossible; starts*/ *-time*] can also be repeated if we add, as further CS, also various dictionary sentences such as *To write is to express one's self*, and then apply the extended ϕ_z to these further CS. For example, if our acceptable sentence had left some un-

---

[18] ϕ_z in 4.2, type 1, zeroes certain constant or determinate words of an operator (in this case ϕ_c producing CS). These words are recoverable from the remainder and add no information. ϕ_z can now be extended to apply to the set of all increments which are recoverable and informationless in the sentence. If we assume the word-repetition conditions for SCS below, then any acceptable SCS which lacks the word repetition must be derived by the zeroing of certain CS which are recoverable from the words of the residue.

[19] ϕ_z in 4.2 has degenerate cases: both *I left him, I being angry* and *I left him, he being angry* zero to *I left him angry*. The extended ϕ_z of the preceding footnote has many more degenerate cases: various alternative $CS \ldots CS$ may be informationless and zeroable after $S_1 CS_2$. But each of these $S_1 CS_2 CS \ldots CS$ has the same acceptability ordering as $S_1 CS_2$.

repeated words, e.g., in

> *He wrote poetry because it was Tuesday, and on Wednesday he had the poetry class.*

we could add, without loss of acceptability,

> *and Wednesday, of course, is the day after Tuesday.*

We can now say that a regular (transformational) $S_1 CS_2 CS \ldots CS$, i.e., one whose acceptability is not reduced from that of S_1, S_2, is one in which each word (except certain constants of the CS) occurs in more than one S; or it is a transform of a regular $S_1 CS_2 \ldots CS$. Regular $S_1 CS_2 \ldots CS$ result from $\phi_{c'}$, in distinction to the simple ϕ_c of 4.2. For arbitrary S_1:

> unary $\phi_{c'} : S_1 \to$ regular $S_1 CS_2 CS \ldots CS$, with a wide range of S_2 determined by S_1 and the particular C;

or   binary $\phi_{c'} : S_1, S_2 \to$ regular $S_1 CS_2 CS \ldots CS$, for arbitrary S_2;

where all $CS \ldots CS$, in both cases, are within a domain determined by the word-repetition condition. Since the added CS are determined (not uniquely) and often zeroable, this more complicated $\phi_{c'}$ can for many purposes be treated as the simple ϕ_c of 4.2, in respect to commutativity, associativity, etc.

5.6.3. *Segment count for reference*

The restrictions due to C have been formulated in terms of the conditions required for regular $SCS \ldots CS$, and the derivability of shorter acceptable $SCS \ldots CS$ from the regular ones is carried out with the aid of an extension of ϕ_z. We now consider the fact that the language contains metasentences about sentences, and note that such metasentences about S_1 may be among the CS which are adjoined to S_1 (and then zeroed). This possibility of adjoining to S_1 a metasentence of S_1 has wide ramifications (5.6.4). Taken together with the linearity of discourse, it makes it possible to speak in the adjoined metasentence of S_1 about the individual segments of S_1.

We thus have an instrument for the counting and classifying of sentence segments. Every sentence is a linear ordering of phonemes and morphemes (at least in respect to their starting point), and also of larger elements.[20] This means that every sentence is a sequence of distinctive sound segments, various subsequences of which are counted, i.e., associated with the natural numbers (as being the first, second, etc., phoneme; the first, second, etc., morpheme; and so on). Also, every sentence is a case of a sequence of

---

[20] The rare case of two phonemes or morphemes which start phonetically at the same point can be treated by a linear ordering of the morphophonemes of a sentence.

word or morpheme classes, since its well-formedness is stated as a class
sequence, so that each word in a proposition is recognized as a member of
a particular class. That is to say, the various segments in a sentence are
classified (a sound classified as being a particular phomene, a word as
being a noun, etc.). Hence each occurrence of a classification in a sentence
is also associated with the natural numbers: a particular subsequence of
sounds in a sentence is the nth case of a given class (vowel, noun, etc.) in
that sentence. It is this counting which makes possible the specifying of
position in cited material, in metatoken sentence (5.4). This pairing of
countings, and of classifications, with the various subsequences in a sentence is equivalent to adding to each sentence S_1 a string of CS, each
added S being of the metatoken form n, A in S_1 is N_{meta} (meaning: the nth
A segment in S_1 is a member of N_{meta}): e.g.,

> S_1, where x is the first phonemic segment and a member of
> phoneme X, and y is the second phonemic segment and a
> member of phoneme $Y, \ldots$, and where $x \ldots z$ is the first
> morpheme and a member of the class noun, ... and so on.

All these added CS are zeroable, as being known to anyone who knows the
language, but their original presence is needed in order to account, by
the known transformations, for such counting and classification references
as: *People like dogs more than the latter do the former.; He meant to say
'Ladies and gentlemen,' but the second noun came ahead of the first.*

These added CS are no longer simple metalinguistic sentences which
merely cite a linguistic entity, such as have been called metatype sentences
(5.4). Here the added CS cites a positioned element within a cited linguistic entity, i.e., the CS cites a pair consisting of the element and its
ordinal number among the like elements within a linguistic entity which
the CS cites. E.g.,

> *People like dogs more than the latter like the former*

can be obtained by means of this counting, together with such added CS
as in the following:

> *Sentence 1, People—first noun—like dogs—later noun—more
> than, Sentence 2, dogs like people,
> where, Sentence 3, 'dogs' in sentence 2 is the same as the later
> noun of sentence 1 and 'people' in sentence 2 is the same as the
> earlier noun of sentence 1.*[21]

[21] In citing, the reference *in sentence 2* can be replaced by *in the sentence 'Dogs like
people,'* and so for *in sentence 1*. But the counted classifier which introduces each segment,
e.g., *Sentence 2* before *dogs like people*, is a count and not a reference, and is not replaceable by the segment which it counts.

These added CS state the occurrence of an element (e.g., later noun) in a linguistic entity (e.g., sentence 1), and to do this a metalinguistic sentence has to cite the linguistic entity and to give the position in it of the element. These are what were called metatoken sentences, in contrast with the metatype sentences.

5.6.4. *Metasentence CS*

Since every grammatical statement is itself a sentence of the language, the above method can be extended so as to provide that any grammatical information about a sentence can be added to that sentence in the form of metatype or metatoken CS.[22] And since the grammatical information is presumed known, in effect, by speakers of the language, these grammatical CS are zeroable. In some cases degeneracies result, with the effect of "abus de langage": Different grammatical statements may apply to the same word sequence, or even to the same proposition; and when the grammatical CS are zeroed, the remaining sentence, or a segment of it, may be taken in more than one way, e.g., as being a relation, or as being the name of a relation.

What is characteristic for language is that particular C and S are grammatically available, which can adjoin to any SCS the metalinguistic and contextual material required to make that $S_1 CS_2$ into a regular $S_1 CS_2 CS \ldots CS$ (and which are then in many cases zeroable). The fact that SCS is regularized simply by further CS and that these include the metalinguistic CS makes the set of sentences into a self-sufficient universe requiring no prior science. For every S_1 can be presumed to be derived by ϕ_z from $S_1 CS \ldots CS$ where the added CS contain all the grammatical, dictionary, and other information necessary for a characterization of S_1 and for relating it to finitely many classifications of the other sentences of the language. The dictionary sentences themselves may be, at least in part, summary sentences for metalinguistic (grammatical) sentences which give the acceptability ordering of a word in respect to other words in each elementary sentence form or operator in which it appears. These added CS would also contain information about the nonlinguistic context in which S_1 is said, if S_1 is not understandable without this information. All these CS, both linguistic and contextual, are zeroable if known to speaker and hearer. Indeed, if they are not known, S_1 alone is not understandable.

It follows from the above that $S_{1_p} CS_{2_q}$, where S_{ix} is a sentence S_i containing a word x, and where p and q have certain dictionary-stateable

---

[22] The same applies to all sentences which state the context in which a given sentence is said, and any other seemingly extragrammatical information necessary to its understanding or acceptance.

relations (such as q being local synonym, or antonym, or classifier of p), can be made regular by reconstructing:

$$S_{1p} CS_{2q} \xleftarrow{\phi_z} S_{1p} CS_{2q} CS_{dict(pq)}$$

where $S_{dict(pq)}$ is a dictionary sentence stating the relation between p and q. More generally, any S_{1q} can be derived:

$$S_{1q} CS_{dict(q)} \xrightarrow{\phi_z} S_{1q}$$

where $S_{dict(q)}$ is a dictionary definition of q, which is zeroable because presumed known. But we can also derive S_{1q} thus:

$$S_{1p} \xrightarrow{\phi_m} S_{1q} CS_{repl(pq)} \xrightarrow{\phi_z} S_{1q}$$

Here S_{1p} is an otherwise identical S_1 containing p instead of q, and $S_{repl(pq)}$ is a dictionary sentence stating that q is replaceable for p (as a local synonym or classifier).

The ϕ_m here simply gives an equivalent form—changing p to q and explaining the change away by the dictionary statement of the local replaceability between them—and then the dictionary sentence is zeroed in the presence of S_{1q} although it would not be zeroable after S_{1p}. In respect to dictionary statements, sentences which differ only in having local synonyms of each other, or classifiers, can come out to be such transforms of each other.

This relation can be used in the opposite direction, for the stating of synonymities. If we can show that all sentences which differ only in having q instead of p can replace each other in all discourses without any change in the acceptability (or meaning) of the discourse, we can reconstruct a sentence $S_{repl(pq)}$.

The synonymity relations among the various C themselves are particularly hard to determine, since the neighborhood in which replaceability has to be tested is not merely the immediate sentence. However, there are clearly fewer independent members of ϕ_c than there are conjunction words in the language. Careful investigation of the similarity requirements on the sentences connected by particular C, and of the discourse neighborhoods in which the C occur, should make it possible to say wherein the various C differ, which are synonymous, etc. It turns out that the *and*, *or* of logic are indeed independent; but it seems to be the case that *but* is a synonym of *and* plus the negative of a certain acceptability condition, that *because* differs in grammatically stateable way from the *implies* of logic, and so on.

5.7. Reference

5.7.1. Citation with position

In 5.6.3–4 the ϕ_z on constants (4.2.2.6, type 1) was extended. Here we will consider an extension of the ϕ_z on repetition (4.2.2.6, type 2), but will find instead a method for using this ϕ_z without extension. The problem concerns the referential semantic effect of this ϕ_z. In 4.2.2.6 it was seen that words in certain secondary positions of a derived (nonelementary) sentence could or must be pronouned or zeroed. This gave each such word a reduced physical shape (phonemic composition), since the phonemes that occupied its position in the sentence were now those of the pronoun or zero. But information is not lost, except for ambiguity being brought in, because the presence of these reduced shapes in a position p tells us that the word at p is the same as the word at a stated other "antecedent" position (or one of several stated positions) in an operand or operator identified with respect to p. Furthermore, in the case of the major occurrences of N, the reduced shape (i.e., the trace of ϕ_z) at p tells us additionally that the N at p referred to the same individual as the N in the antecedent position.

We will consider here precisely what syntactic relations make it possible to identify the antecedent (5.7.1) and to refer to sameness of individual (5.7.2), and what extension may be needed to carry ϕ_z beyond the scope of a sentence (5.7.1).

The possibility of referring to a particular occurrence of a word in a position in a sentence is due to the counting and grammatical classification which is associated with each segment, and which is reported in meta-token sentences (5.6.3). The fact that two positions in a sentence or cited segment are filled by occurrences of the same word, and that the positions are corresponding or related ones in the operands and operators of their respective sentences, is then expressed by a metasentence which includes the necessary metatoken sentences and which can be viewed as being attached to the sentence which it discusses (5.6.4). Thus, given (1), (2) below, we could assume sources such as (3), (4) as their respective sources:

(1) *I promised to come.* ← *I promised that I would come.*
(2) *I will go if you will.* ← *I will go if you will go.*
(3) *1, I promised that I would come, where 2, 'I promised that' is ϕ_s[23] and 'I will come' is its operand, and 3, the subject (or: first noun) of the operand is the same as that of the ϕ_s.*

---

[23] In a subclass of ϕ_s that changes the tense of the operand to *would*.

(4) *1, I will go if 2, you will go, where 3, 'if' is in C_s, and 4, the verb of S_2 is the same verb as in S_1.*

In source sentences constructed like (3) and (4), a morphophomemic operator ϕ_m would then replace the last metasentence *CS* (in conjunction with the information in the preceding metasentence *CS*) by a zeroing or pro-wording of the positioned word in the sentence numbered 1 in (3) and 2 in (4). We thus obtain (with a known morphophonemic change between *would* and *to*):

> *I promised to come.*
> *I will go if you will.*

In general, then, the possibility of cross-reference in a sentence is obtained from the ability of metalinguistic sentences to cite the ordinal number of a component sentence in an *SCS ... CS*, and also each relevant segment of each *S*, thus identifying an individual occurrence of the sentence (and of its parts) in that *SCS ... CS*. Just as all sentences are linear orderings of segments, so all *SCS ... CS* and discourses (see 5.8) are linear orderings of sentences (or at least of their beginnings). Thus

$$S_i CS_j \ldots CS_q$$

is more precisely:

$$1, S_i C2, S_j \ldots Cn, S_q$$

or:

> *Sentence 1, S_i C sentence 2, S_j ... C sentence n, S_q.*

We now ask about cross-reference beyond the limits of a sentence. First, we note that there is no cross-reference between discourses, except by explicit or implicit citing of one discourse within another. The material in different discourses is not linearly ordered. But the material within a discourse is linearly ordered. The possibility of cross-reference depends upon stating the position of linearly ordered objects. Many zeroings and certain pronounings (e.g., with *wh-*) occur only in respect to antecedents within the same sentence. However, certain pronounings (and, limitedly, a few zeroings) occur in respect to antecedents in other sentences of the same discourse, almost always preceding sentences. How is the antecedent identified? We have seen that when the antecedent and the pronouning (or zeroing) are in the same sentence *A*, the identification is carried out by metatoken sentences *M* conjoined to *A* and containing the addresses of both the antecedent and the pronoun (or zero) in A, as in (3), (4) above. If now the metatoken sentence M which refers to the pronoun (or zero) in A is to contain the address of some antecedent in another sentence B, it must state the discourse position of B. M itself must be adjoined to A, otherwise it will have to refer to A too, by stating its position also. To state the position of B, M has to cite the whole preceding discourse, giving the position of B in it; otherwise we would have to identify the discourse (which contains A, B and M) by some word like *this*—which is precisely of the class of pronouns which we are here defining. A linguistically more interesting alternative is to conjoin (a repetition of) the antecedent-bearing sentence B into the pronoun-bearing A. The identifying of an antecedent in another sentence (always of the same discourse) has thus been reduced to the previous case of identifying an antecedent in the same sentence. The question of what sentence is the antecedent-bearing one, and how far in the discourse an antecedent can be, now becomes the question of what sentences of the preceding discourse can be conjoined into A, while maintaining a reasonable discourse and satisfying the *SCS* requirements for A.

Hence to any sequence of n sentences $S_1 C S_2 \ldots C S_n$ (where C now indicates either conjunction or period) we can add an $n + 1^{\text{th}}$ sentence which can cite the whole preceding discourse $S_1 C S_2 \ldots C S_n$ and say something about any of the ordered sentences and segments in it. If in a discourse, a sentence S_n contains a cross-reference R (zero or pro-word) to some word W in some position x of the discourse, then S_n is derivable from S_n', a sentence identical with S_n except for containing W in place R, plus a metasentence S_{n+1} (usually following S_n) which cites enough of the discourse to contain x and S_n, and states the sameness of the two occurrences of W and the grammatical relation between their positions in their respective sentences.[24] A pro-word or zero cannot refer to any material except such as is cited in the sentence which states the grounds for the pro-wording or zeroing. This means that it cannot refer to material in the preceding or other discourses without citing the preceding or other discourse. But when it refers to preceding material in its own discourse it may use the sentence numbering instead of the full citation; i.e., the citing of the preceding discourse is zeroed because it is recoverable from the preceding discourse.[25]

[24] The required relation between the two positions depends on the operators: ϕ_s, particular ϕ_c, etc. (4.2.2.6).

[25] In the example in 5.7.2, the citing of the preceding discourse, namely, *In 1, SC2, S...Cn* − 2, $S_i C n − 1$, S_j, can thus be zeroed, leaving '$n − 1$', etc., to refer implicitly to a section in the preceding discourse.

5.7.2. Reference to individual

It remains to see how the activity of zeroing or pronouning is carried out. If the zeroing is automatic in a given operator combination (as in *The carton is smaller than the book*, from ϕ_z ... *The book is small*), we try to make it a morphophonemic part of that operator combination. If the zeroing or pronouning is optional over a domain defined purely in terms of words and positions (as in *He went and she went, He and she went*), it is simply the operator ϕ_z, although we might want a metasentence announcing the sameness of word, as in (3), (4) of 5.7.1. But if the zeroing or pronouning depends upon anything beyond such grammatical information, then there must have been some sentence B which gave this additional information, and whose relevance to the pronoun-bearing A is provided by B being conjoined to A together with the metatoken sentences discussed above. This is the case with pronouning in the major occurrences of N, where pronouning occurs not simply when an N has been repeated but only if the N refers to the same individual as the antecedent: e.g., for the *wh*-pronouns (*who, which*, etc.) and for kernel-sentence nouns which are counted or identified (preceded by *the*) in the two occurrences.[26] This can be seen in *A man who borrowed a book left a note*, where both actions are by the same man. If they are by different men, we have only *A man borrowed a book and a man left a note*.

We have seen that language can be described simply as sequences of classified words. The words (or at least subsequences of them) have semantic interpretation, and we can say that they designate classes of objects and relations and events in the real world. Words do not in general designate uniquely each individual object in the world, although a particular occurrence of a word may. The sameness of the man in the sentence above is not guaranteed by some word which names that unique person. We therefore ask how the semantic effect of sameness of individuals is obtained from sequences of words which by themselves do not have such a meaning.

The requirement of sameness of individual can be obtained from the metasentence *CS* which is the basis for the morphophonemic operator of 5.7.1. If the source sentence does not contain, in its metasentence *CS*, the assertion about the sameness of individual, the zeroing or pronouning of counted nouns does not take place. If it does contain such an assertion, then the information about the sameness can be expressed by the zeroing or pronouning instead of by the assertion. Thus the *wh*-words, which connect two sentences containing the same noun N_i require that the two

---

[26] Hence not a transformationally derived nominalization, like *truth, arrival*, etc.

occurrences of N_i in the two sentences refer to the same individual. This can be expressed with the methods of 5.6 if we can show (as can indeed be done) that *wh-* can be replaced by an existing other C, provided that a certain metatoken CS is added:

$$n - 2,\ S_{i(a, N_1)}\ \textit{wh-}\ n - 1,\ S_{j(b, N_1)}\ \overset{\phi_m}{\longleftarrow}\ n - 2,\ S_i\ \text{and}\ n - 1,\ S_j$$
and n, *In* '1, $SC2$, $S \ldots Cn - 2$, $S_i Cn - 1$, S_j' *the pair* b, N_1 *in sentence* $n - 1$ *refers to the same individual as the pair* a, N_1 *in sentence* $n - 2$.[27]

Here $S_{i(a, N_1)}$ and $S_{j(b, N_1)}$ are sentences containing a particular N_1 in positions a and b of S_i and S_j, respectively.

E.g., *A book which I bought has disappeared*
   $= [1,\ A\ book\ has\ disappeared]$ *wh-* $[2,\ I\ bought\ a\ book]$
   $\leftarrow [1,\ A\ book\ has\ disappeared]$ *and* $[2,\ I\ bought\ a\ book]$
   and $[3,\ In\ '1,\ a\ book\ has\ disappeared\ and\ 2,\ I\ bought\ a\ book',$
   'book' in post-verb position of sentence 2 refers to the same individual as 'book' in pre-verb position of sentence 1].

There is evidence suggesting that " an individual" can be replaced in certain language contexts by "counted in the same counting act." The situation is as follows: A word, e.g., *man* means a certain class of objects or relations. When we adjoin a number (including the word *a*) to an occurrence of a noun, the effect is that certain members of the class have been paired with the natural numbers, up to the number which has been adjoined: *a man, three men*; we will call this a counted noun. Now it appears that in certain forms, zeroing and definite pronouning mean "the same individual" only when they are carried out on a counted noun (including proper names used for specified individuals) and in respect to an antecedent counted noun consisting of the same words. Thus

A man came and a man left.

may or may not refer to the same individual; imagine that the speaker adds: "That is all I could see; I couldn't tell if it was the same man."

---

[27] Here and in the next paragraph *the* is a constant and could be replaced by its local synonyms or transformational sources; there is no *a same individual* as distinct from *the same individual*. Since reference specifies same individual only in the case of material objects (which are the counted nouns), we can replace *X is same individual as Y* by *the space-time coordinates of X equal those of Y*, or the like. Then "same" is avoided in favor of "equal." For inclusion in the set of prime sentences (4.3.3), a fixed form would have to be chosen for this *wh-* carrier, the discourse position of the sentences referred to being given as distances from the *wh-* carrier.

But the zeroing in

A man came and left

refers only to the same individual. In

Men came and left ← Men came and men left

the zeroing does not mean that the men must have been the same. But in

Two men came and left ← Two men came and two men left.

the zeroing means that the individuals were the same (although they may or may not have been the same in the unzeroed form, precisely as for *a man*). In uncounted nouns, as in other word classes other than nouns, for which there is no counting, zeroing does not mean same individual:

Anger is ineffective and destructive. ← Anger is ineffective
and anger is destructive.

Similarly, *wh*- does not individuate when the entity which it pronouns is uncounted or uncountable, as in:

His lying, which I dislike, is a problem.

The ability to refer to the same individual depends here apparently not on an absolute identification of the individual, but on an identification of a particular counting act on a word.

Similar methods give a source for *the* and the various definite pro-adjectives (*this*, etc.). For various uses of *the*, we can give various sources, such as the following:

$$n - 1, S_{1(the\ N_1)} \xleftarrow{\phi_m} n - 1, S_{1(N_1)} \quad wh\text{-}\ n,\ N_1\ is\ same\ as$$
$$N_1\ before\ n - 1,\ S_1.[28]$$

E.g., *He bought the book ← He bought a book which is identical with a*
book recently mentioned before 'He bought
the book.'[29]

---

[28] The added *wh*- *S* can in turn be decomposed as in the preceding paragraph. If N_1 was not mentioned before but is understood (e.g., as being the book that the speaker is interested in), then the fact of its being otherwise identifiable is added as a *CS* before $S_{1(N_1)}$, and becomes the "recently mentioned" referred to. Cf. Beverly Robbins, *The Definite Article in English Transformations*. The Hague, 1968. (Papers on Formal Linguistics, No. 4.)

[29] Of course, *recently* can be replaced by citing a near position in the preceding discourse.

Similar methods also give a source for the pronouns.

E.g., $n - 1, S_{1(p,he)} \leftarrow n - 1, S_{1(p,N_1)}$ *wh- n, N_1 is, in the sentences before $n - 1$, S_1, a (usually, the most) recently mentioned human, masculine, singular noun or the most recent one in a position corresponding to position p in S_1.*[30]

It is of course not the intention here to claim that the proposed source *CS* are really used in the language. Their formulation shows only that the ability of pro-words and zeroing to mean sameness of individual can be expressed and replaced in the affected sentences themselves by added *CS* whose only peculiarity is that they refer to earlier parts of the discourse in which they occur.[31] Sameness of individual is therefore not expressed by primitive terms having the novel power of reference, but by utilization of the linearity inherent in every discourse.

5.7.3. Impredicatives

We have seen that pro-words can refer only to preceding or immediately following parts of the discourse, since the referents are defined by occurrence (position-and-word pairings) in material which is cited in the metalinguistic *CS*.[32] This makes it possible to distinguish certain impredicative sentences, which are at the root of logical paradoxes, from the grammar of the rest of the language. These impredicative sentences are the ones which

---

[30] This can be stated more precisely, but the statement involved. One can try to state what operators (and how many of each) can intervene between *he* and its antecedent, and what similarities of neighborhood (in terms of ϕ, or the f of 7.1.2) will make a given N_1 occurrence likely to $\rightarrow$ *he* in respect to another (antecedent) N_1 occurrence.

[31] More explicitly, to counting in the domain of nouns in certain positions in the preceding portion of the discourse.

[32] This explains, for example, why $\nexists$ *He will go, if Paul can.* The only forms that exist are:
(1) *Paul will go, if he can* $\xleftarrow{\phi z}$ (2) *Paul will go if Paul can and the two Pauls are the same.*

If he can, Paul will go $\xleftarrow{\phi p}$ (1)

If Paul can, he will go $\xleftarrow{\phi z}$ *If Paul can, Paul will go and the two Pauls are the same* $\xleftarrow{\phi p}$ (2).

Such cases as stories that begin with *He* have *he* as indefinite *someone* rather than as definite pronoun. There remain certain stylistic uses (*He will do it, Paul will*), which can be separately treated. All this is not to say that a sentence cannot refer to a later sentence in the discourse; only that pro-words cannot in general be used for this. There are a few aberrant cases such as a pro-word for an immediately following operand: *I want to say this: You are wrong. ← I want to say: You are wrong.*

can be reduced to a form containing self-referring pronouns, as in:

This sentence is false.

(or, for that matter, *This sentence contains five words* or *This sentence contains four words*). According to the discussion above, the grammatical source for this would have to be:

$n - 1$, *A sentence is false. wh- n, 'sentence' in sentence* $n - 1$ *is the same as a recently mentioned 'sentence' before sentence* $n - 1$.

But this refers to a preceding mention of "sentence," if any, and has no relation to the impredicative. The transformational source of *this* and the like in the nth sentence does not provide for referring to the nth sentence, because the nth sentence cannot cite itself, since the cited material, being the segment X of an X *is* N_n kernel sentence contained in the nth sentence, must be completed before the nth sentence can be completed grammatically as a sentence.

One could, of course, say that there is another interpretation of *this*, apart from reference to antecedent, as referring to the ongoing activity. When such *this* is used for nonlanguage events (e.g., *This concert is beautiful*), we can prefix a sentence about this event, as we would for any *this* whose antecedent did not appear (e.g., *I like this picture* ← *We are looking at a picture and I like this picture*); then the deictic *this* of pointing is reduced to the antecedent *this*. But when *this* is used for ongoing language material, we cannot prefix a sentence providing the material; and an *ad hoc* definition would be needed for such uses of *this*.[33]

5.8. *Discourse*

All occurrences of language are discourses, each being segmentable (3.6) into stretches of sound whose structure can be characterized as some connected segment of a sentence. But discourses are not merely sequences of sentences (including of *SCS ... CS* forms); they show a certain additional structure.[34]

Discourses are all the connected occurrences of language, from start to finish: comments, articles, conversations, etc. It can be shown that in

[33] There are intermediate situations. If a sentence begins *In this paper* or *In the present paper*, the paper as a whole may be looked upon as an outside object which exists even if the sentence is never finished or ceases to be grammatical. But for *In this sentence*, the sentence may be meaningless unless it is finished in particular ways.

[34] Short lone sentences do not show the structure described below, but can be said to satisfy it vacuously.

each discourse there are certain classes of segments which recur in some compactly characterizable way. These segments are not whole sentences but constituents of the sentences of the discourse, more precisely constituents of transforms of the sentences. To discover, for each discourse, what are its recurring segments, we use an equivalence relation (transitive except for subscript, symmetric, reflexive) on morpheme sequences, recursively defined as follows:

$$a = {}_0 b . \equiv . a \text{ is the same morpheme sequence as } b^{35}$$
$$a = {}_n b . \equiv . env \; a = {}_{n-1} env \; b$$

where $a, b, \ldots$ are morpheme sequences, and $env \; a$ is the remainder of the sentence which contains a; that is, $env \; a$ is the sentential environment or neighborhood of a, and is itself a (possibly broken) sequence of morphemes. ($env \; a = {}_{n-1} env \; b$ is taken to mean that at least some part of $env \; a = {}_{n-1}$ the corresponding part of $env \; b$, and that any other parts of $env \; a = {}_{m<n-1}$ the corresponding parts of $env \; b$. That is, $n - 1$ is the highest subscript of equivalence between any part of $env \; a$ and the corresponding part of $env \; b$.)

The equivalence $a = {}_0 b$ is used only when we can find a chain of equivalence with ascending subscripts. To find such an ascending chain, it is usually necessary that a and b occur in corresponding grammatical positions within their respective sentences, or within the transforms of their respective sentences. Ubiquitous words like *the, in, is* will usually not satisfy this condition, and are therefore useless as a base for a chain of equivalences.

The equivalence $c = {}_n d$ between particular morpheme sequences may be reached by more than one chain. The degree n of the equivalence between them will be understood to be the lowest subscript of $c = d$ in any chain in which $c = d$ appears.

Because of the grammatical variety which is irrelevant to the content of the discourse, it is difficult to find segments which have the same sentential environment. However, the sentences of the discourse can be

---

[35] After the equivalence classes have been set up and their relative occurrence studied, we may find reason to say that in a particular case $a \neq {}_0 a$: i.e., that a particular occurrence of the morpheme sequence a is not discourse equivalent to the other occurrences of the morpheme sequence a. This will happen if we find that accepting the equivalence in this particular case forces us to equate two equivalence classes whose difference of distribution in the double array described below has reasonable interpretation. Such situations are rare; and in any case the equivalence chain has to start with the hypothesis that occurrences of the same morpheme sequence are equivalent to each other in degree zero. The equivalence relation then states that a is equivalent to b if $\exists n, n \geq 0$, such that $a = {}_n b$ by the formula given here.

transformed paraphrastically (i.e., by ϕ_p, ϕ_z, ϕ_m, or certain $\phi \ldots \phi$ products which do not change the informational content of S, or by the inverses of these), and this can be done in such a way as to maximize the membership of the equivalence classes. For a given discourse D_i we will call any discourse TD_i, which is obtained from D_i only by paraphrastic transformations on the sentences of D_i, a transform of D_i. For each D_i there are one or more optimal TD_i, in which there exists an equivalence chain longer than in any other TD_i. The equivalences found in the optimal (or any) TD_i can be considered valid for D_i, since TD_i is a paraphrase of D_i.

In establishing equivalence chains for a discourse, we may use ancillary sentence reconstructions of the types presented in 5.6–7. For articles in a particular science, any definitions and synonymities of the science sublanguage, or any relevant statements which are well known and acceptable to all scientists in the field, can be adjoined as CS to the article. In general, the context in which a discourse is stated and understood can always be adjoined to that discourse as additional CS.

This method necessarily divides the discourse into recurring sequences of certain equivalence classes, and thus constitutes a grammar of the particular discourse. For a sketch of how discourse analysis works, we take the last paragraph (sentences 21–26) of an analyzed article.

21   The presence of the oxidized A-chain of insulin in Group B was
22   not expected. In fact it was hoped that the A-chain would exist as
      an α-helix in solution and would therefore serve as a model substance
23   of known structure in the study of denaturation. Instead it was found
      that the A-chain possesses rotatory properties which resemble those
      of clupein very closely, but do not resemble those of insulin itself.
24   Most striking is the fact that the specific rotations of clupein and the
      A-chain are virtually unaffected by strong solutions of urea and
25   guanidine chloride. Ordinary proteins, including insulin, undergo
      changes in specific rotations of 100 to 300 under these conditions.
26   These results suggest that the oxidized A-chain is largely unfolded in
      aqueous solution and are in agreement with the recent finding that
      the peptide hydrogen atoms of the A-chain exchange readily with
      D_2O, whereas those of insulin do not.

In a transform, TD, in which the 27 sentences of this article, D, are transformed into 82 sentences of TD, the above paragraph is analyzed as shown on pages 150 and 151.[36]

[36] For the whole article, see Z. Harris, *Discourse Analysis Reprints*, Papers on Formal Linguistics 2 (1963) p. 20ff.

Column 1 gives the sentence number in D, column 2 gives the sentence number in TD. Each following column contains an equivalence class; i.e., the entries in each column have been shown to be equivalent to each other by the operation described above. It turns out that throughout the whole article, every sentence of TD consists of the $HRLK$ equivalence classes (or of two or three out of these four), with intervening verb equivalence classes.

The interpretation is, of course, not that all members of an equivalence class are synonyms of each other, but that the difference for the given discourse between any two members x_1, x_2 of one equivalence class X corresponds to the difference between the corresponding members y_1, y_2 of the class Y with which they occur. I.e., the discourse contrasts x_1, y_1 with x_2, y_2 or equivalently, x_1, x_2 with y_1, y_2.

It is possible, in a discourse, to collect all the sentences which have the same classes, so as to see how the members of a class change correspondingly to the change in members of the other classes, and so as to compare these sentences with sentences having in part other classes. In many cases it is possible to obtain, by formally definable although as yet only tentative operations, a summary of the argument; and it is possible to attempt critiques of content or argument based upon this tabulation of the structure of the discourse.

If we consider many discourses of the same kind, we find that their analyses are partially similar. For example, in scientific articles it is characteristic that we obtain a set of sentences in the object-language of the science, and on some of them a set of metascience ϕ_s and ϕ_c operators which contain words about the actions of the scientist, or of prior sciences (e.g., logical relations). In the table the last column (and some of the C column) is of this nature.[37]

The result of discourse analysis is essentially a double array, each row being a sentence of the TD and each column a class of sentence segments which are equivalent by the discourse operation. All further analysis and critique of the discourse are based on the relations contained in the double array. The discourse dependence of later sentences in a paragraph upon the first sentence is different from the dependence of S_2 upon S_1 in $S_1 C S_2$. In a discourse, the later sentences are not independent, but we can extract from them certain factors of modification which operate on segments of the first sentence to produce segments of the later sentences. These factors are independent of the first sentence, and it is these factors then that constitute the second dimension. It is in this respect that discourse, as a

[37] In TD sentence 65, a metascience entry should be extracted from the word-set *model...known...serve...study*.

1	2	3	4 H	5	6 R	7
21	61		the oxidized A-chain of insulin			is presen in
	62					
22	63	In fact				
	64		the A-chain	as	an α-helix	
	65	and therefore	the A-chain as model substance	of	known structure	
23	66	Instead				
	67		the A-chain	possesses	rotatory properties	
	68		clupein	possesses	rotatory properties	
	69	but	insulin itself	possesses	rotatory properties	
24	70					
	71		clupein	has	specific rotations	
	72	and	clupein	has	specific rotations	
	73	and	the A-chain	has	specific rotations	
	74	and	the A-chain	has	specific rotations	
25	75		ordinary proteins	have	specific rotations	
	76	including	insulin	has	specific rotations	
26	77					
	78		the oxidized A-chain	is	largely unfolded	
	79	and				
	80		the A-chain	has	peptide hydr. atoms	
	81	whereas	insulin	has	peptide hydr. atoms	
27	82					

8 L	9	10 K	11 METASCIENCE
Group B			
			61 was not expected
	would exist in would serve in	solution the study of denaturation	64–5 was hoped
			67–9 was found resembling X very closely resembling X very closely not resembling X
	virtually unaffected by virtually unaffected by virtually unaffected by virtually unaffected by	strong solution of urea strong solution of guanidine chloride strong solution of urea strong solution of guanidine chloride	The fact 71–4 is most striking
	undergoing changes of 100% to 300% under undergoing changes of 100% to 300% under	these conditions these conditions	
	in exchanging readily with not exchanging readily with	aqueous solution D_2O D_2O	Results 71–6 suggest 78 results 71–6 are in agreement with recent finding 80–1
			A detailed report of this work will appear later

two-dimensional structure, differs from language which is a one-dimensional structure even in its $SCS \ldots CS$ word-repetition requirement.

5.9. Sublanguages

5.9.1. Sublanguages of sciences

Certain proper subsets of the sentences of a language may be closed under some or all of the operations defined in the language, and thus constitute a sublanguage of it. This holds for such subsets as the sentences of a particular phonetic dialect, and also for stylistic subsets: e.g., if S_i is in a particular style, ϕS_i is also in that style. A more interesting case is that of a subset of sentences which satisfy some grammatical conditions not satisfied by the language as a whole. An example is the metalanguage. Each sentence of a grammar (although not necessarily each sentence of a grammatical discussion as actually written) says something about sentences of the language, or their parts, or classes of these; hence each contains (or is derived from a sentence which contained) one of those members of N_n that names these objects (*word*, *sentence*, etc.). When these particular words are taken as the extension of the new subclass N_{meta} of N_n, we obtain the grammar as a sublanguage consisting of all those sentences that contain N_{meta}. In the language as a whole (including the grammar), this subclass has no occasion to be recognized as a separate entity. By this means, we can say that the grammar of the metalanguage is characterized by a certain grammatical property which the language as a whole does not satisfy.[38]

This situation appears characteristically in the language of various sciences, i.e., in sets of sentences devoted to describing (correctly or incorrectly) particular areas of structured phenomena. Here all the sentences satisfy certain grammatical restrictions which do not hold for the language as a whole. For example, (1) *The polypeptides were washed in hydrochloric acid, The proteins were treated with acid*, are sentences in biochemical language. (2) *Hydrochloric acid was washed in polypeptides* is not. It is not a question of truth, since (1) may also be not true, if in point of fact the polypeptides were not washed in HCl. The excluded sentences are more similar to what would be called nonsense than to falsity, except that they

---

[38] The mere fact of requiring each sentence of the metalanguage to contain an X *is* N_n kernel sentence is itself a rule of the grammar of the metalanguage which does not hold for the grammar of the whole language (or for the grammar of the object-language alone).

[e.g., (2)] are more definitely unacceptable in science discourses than is nonsense in the language as a whole. In the language as a whole no such sharp unacceptability attaches to (2): it may not even qualify for a grading of " nonsense "; there are sentences with somewhat extended uses of *wash* which come close to it. In describing the sentences of biochemistry we can define particular subclasses of words, such as names of proteins or various classes of molecules, and names of solutions, reagents, etc., and verbs for classes of reaction or laboratory activities that are carried out on the molecules. Of these specially defined subclasses, only particular sequences will be found in (true or false) well-formed sentences in biochemistry discourses. These sentences are also in the language as a whole, but other sentences in the language do not keep to the particular sequences of these particular words, so that the biochemical word subclasses and their well-formed sequences do not exist as such for the language as a whole.[39]

The axiomatic view of grammars is that a grammar constructed for a language (a set of sentences) consists of a set of word and morpheme classes (and subclasses), a set of well-formed sequences of these (elementary sentence structures), and a set of transformational rules which derive one sentence structure from another. In this sense, the grammar of the sentences in a particular science contains items additional to those of the grammar of the language as a whole.

5.9.2. *Sublanguage grammar intersects language grammar*

The possibility of a subset of sentences, such as those in a science or in science discourses in general, having a special grammatical structure while also satisfying the grammar of the language as a whole, is due to the fact that a sentence structure can be taken to be the sequence of subclasses (recognized in the whole language or only in the science sublanguage) to which the successive words of the sentence belong, or equally well the sequence of classes (noun, verb, etc., recognized in the whole language) to which they belong: in the example above, (1) as a sentence of English is N *is Ven in* N; as a sentence in English language biochemistry, included in English, it is N_{mol} *is* V_{sol} *en in* N_{sol}, where the subscripts name particular subclasses defined for biochemistry sentences.

Even the small classes which fill the role of transformational constants, such as prepositions and conjunctions, which have always been considered

---

[39] In addition, in the sentences of a science the local synonyms (6.5) of the words which are members of these subclasses (and of many other words, too) are not the same as the local synonyms of these words in all sentences of the language.

to be unextendable objects in grammar, can receive new members in particular subsets of sentences, thus increasing the grammar for these sentences. The creation of new members of prepositions P and conjunctions C is possible because certain grammatical sequences of morphemes have the same neighbors within a sentence form as do P or C.

For example, in the language as a whole, a ϕ_s operator with following S may be attached by some zeroed C to another S, with the effect that the ϕ_s operator appears between sentences, as though it were a conjunction. Thus

wh-: *I promised with a provision. The provision is that he will promise too.*
$\rightarrow$ *I promised with a provision which is that he will promise too.*

This becomes, by ϕ_z on *which is*: *I promised with the provision that he will promise too*, where *with the provision that* functions like C. More unusually, *I promise, provided he will promise too*, where *provided* functions as C, is derivable by ϕ_z, in the grammar of the whole language, from *I promise, it being provided that he will promise too*, which comes ultimately from ϕ_s: *N provides that he will promise too*. These morpheme sequences can be considered members of C not only because they occur between two sentences but also because the required similarities between the two sentences, or the required neighborhood of further sentences, is precisely of the kind required by C.

It may even be that scientific writing will prove to have more complicated structures, such as ternary conjunctions, i.e., conjunction pairs requiring particular kinds of similarity and differences among the three sentences involved. In any case, it has different discourse structures (5.8) than colloquial or literary texts. Such rules would be established if we show that only certain sentence sequences are found in this writing, more restricted than in the language as a whole.

We have thus seen that sublanguages can exist whose grammar contains additional rules not satisfied by the language as a whole. It also happens that some of the grammatical rules of the language as a whole disappear, i.e., do not apply, in a sublanguage. Since the sublanguage must satisfy the rules for the language, this disappearance is possible only if the rules are satisfied vacuously in the sublanguage, i.e., if certain word classes or well-formed sequences or transformations do not appear in the sublanguage. We can then say that these rules do not apply in the sublanguage. This holds for scientific writing, where various of the grammatical statements for poetic, colloquial, etc., styles are satisfied by default. It holds more explicitly for the sentences of descriptive and theoretical, as against experimental, matter in science. For as was seen (5.8), it is possible by means of purely linguistic methods to separate out a scientific discourse into a set of object-language sentences which talk only about the objects of the science and their interrelations, and a set of metascience operators or sentences which talk of the actions, observations, opinions of the scientist in respect to the object-language sentences (or objects). These portions of scientific texts provide us with a set of object-language sentences, in which many items of the whole language do not appear, e.g., various subclasses of ϕ_s with human subjects (*know*, *hope*, etc.).

Thus the sublanguage grammar contains rules which the language violates and the language grammar contains rules which the sublanguage never meets. It follows that while the sentences of such science object-languages are included in the language as a whole, the grammar of these sublanguages intersects (rather than is included in) the grammar of the language as a whole.

5.9.3. *Conjunctions of sublanguage and language*

Because a conjunction requires (5.6) that the sentence following it have certain similarities to the sentence preceding (or, that particular kinds of sentences be further conjoined to compensate for the excess dissimilarities), each $SCS \ldots CS$ preserves certain properties of its initial S. As a result, certain sublanguages, including the science languages,[40] have, under C_s, a relation to the language as a whole (in respect to all other operations on S) similar to that of a right ideal in a ring. If S_1 is in the sublanguage, and S_2 is not, $S_1 C_s S_2$ retains properties of S_1 and is in the sublanguage. But in some cases, this holds only for those conjunctions which require strong similarities; or else it holds if the special grammatical properties of the sublanguage are defined only on the first S of each of its $SCS \ldots CS$.[41]

The import of this right-ideal type of construction is that certain subject-matter restrictions (referring thus to meaning, as against the material-implication conjoinings of sentences) are determined for every $SCS \ldots CS$ sequence by its first sentence (in the time order in the underlying regular form, i.e., before permutations). The possibility arises of covering the language with a system of such right-ideal-like subsets, whose intersection may be empty or may consist of certain distinguished sublanguages.

---

[40] But not, for example, phonetic dialects.
[41] Many languages have grammatical properties which hold or fail only for the first S of its $SCS \ldots CS$, or for the first S of a discourse: e.g., the rules for pronouning and zeroing.

5.10. Language-pair grammars

The grammars of all languages are not arbitrarily different from each other. The similarities in basic structure are sufficiently great as to make it possible to take the similarly formulated grammars of any two languages and combine them into a single grammar which exhibits the grammar which is common to the two languages and relates to it the features in which the two grammars differ. Explicit and rather obvious methods for doing this can be formulated,[42] and can be so devised as to exhibit the structural difference between any two grammars.[43]

In particular, the types of base transformations of various languages are very similar one to the other. The elementary transformations of a particular type, say ϕ_s, in one language are approximate translations of the corresponding base transformations of the same type in another language, to the extent that appropriate correspondences can be determined. Furthermore, as was seen in 4.4, if a sentence in one language is transformationally decomposed and is also translated into another language, the transformational decomposition of the translation will consist largely of a similar ordering of corresponding transformations.

We have a commutative diagram of translation and decomposition:

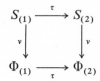

where τ are mappings from the sentences S or prime sentences Φ of language 1 to their translation in language 2, and v is the natural mapping from the set of sentences onto the set of transformational traces (i.e., products of prime sentences). In the translation of Φ, provision has to be made for differences between the two Φ (i.e., K, ϕ) systems as evaluated by the language-pair method mentioned above.

This means that in principle language translation can be effected by decomposing sentences in one language, translating the components (elementary sentences and transformational instructions) into another language, and recomposing the sentence in the other language. Word meaning-ranges correspond (hence translate) better in Φ than in S.

[42] Z. Harris, Transfer Grammar, *International Journal of American Linguistics*, 20 (1954) 259–270.

[43] If grammars can be constructed for particular science sublanguages (5.9), one might conjecture that subsciences which are abstractly equivalent should prove by this language-pair method to have essentially the same grammars.

6

Regularization beyond language

6.0.

We have seen what sets and operations characterize language and give the interesting relations within it. We want to obtain out of this a general statement of what are the essential relations of language, a statement sufficiently general as to determine an abstract system any interpretation of which would be a natural language or its equivalent. The immediate characterization of a natural language always contains some elements, relations, or restrictions, that violate a complete regularity, or that hide a more interesting regularity than appears on first analysis. This is due to various features of language. One is the existence of automatic operations which regularize one kind of structural property at the expense of another: e.g., the verb *be* (6.6) which regularizes word-class sequences ($NtV\Omega$ sentences) but confuses predicates (sentence-making operators, like *sleeps* in *He sleeps*, and *man* in *He is a man*) with verb objects (second members of arguments, like *man* in *He saw a man*). Another is the fact that some of the most important regularities of language are stated not on readily identifiable entities but on complexly defined objects of which the readily identifiable entities are components or complementary variants. This may be due to the fact that the development of language is in part a regularization of previously irregular or differently-regular relations. An example of this is the system of local synonymity in 6.5. A third feature is that for various reasons having to do with language change (including the preservation of isolated cases of no-longer-used forms, and the new analogizings) there always exist some unique and exceptional forms.

There are, thus, particular sources for the existence of partial regularities in language, and also for the existence of further regularities underlying these. It follows from such considerations that not only are the immediately observable objects of language (e.g., sounds, pauses) not the best elements for a scientific description of it, but even the elements set up in linguistic science (phonemes, morphemes, sentence forms) are not the best sets on which an abstract system for language can be formulated.

One can try to remove restrictions, to increase the regularity of certain sets of forms, or to generalize the operations. In some cases one regularization may conflict with another, as in the case of *be* above. For example, there are so many derivations in both directions between ϕ_v and ϕ_s that one can redefine a modified ϕ_v so that all ϕ_v are derived from ϕ_s; or one can redefine a modified ϕ_s so that all ϕ_s are derived from ϕ_v (6.1). Finally, it may appear that in some respects the regularizing of language structure can proceed in more than one direction, and additional (interpretational) considerations must be brought to bear (6.8).

Regularizing the grammar without changing the set of sentences which the grammar describes means replacing a grammatical or dictionary difference by a morphophonemic operation. If ϕ_i operates on a restricted domain X and ϕ_j has the same effect as ϕ_i but operates on domain Y, we define a ϕ_m which operates on the pair ϕ_i, X, such that $\phi_m \phi_i = \phi_j$. Regularizing the grammar by extending the set of sentences to include nonextant (source and intermediate) sentences which are implicit in the transformational structure of the extant ones is different, but does not change appreciably the informational capacities of the language.

We will therefore sketch below some of the major possibilities of regularizing and generalizing the grammars of languages, without changing the universe to be either more or less than the natural language in question. The details of such regularization are of interest only to linguistics; but the brief sketch below will be relevant to the formulation of an abstract system in Chapter 7.

6.1.   Degrees of freedom of elements

The work in linguistics, as in comparable material of any science, has been to determine what are the independent elements, regular combinations of which characterize all more or less independently occurring entities—in this case, sentences. In the attempt to construct freely-combining independent elements out of what appeared to be dependent elements which entered only into restricted combinations, more powerful results can sometimes be obtained by extending the degrees of freedom (of sequencing with other classes) in which a set of elements is defined.

An example from phonology: When the sounds of each utterance are cut into successive segments, it is possible to determine maximally independent segments; these are then called occurrences of phonemes (or letters) of the language. The phonemes occur successively, and have no length defined. However, we find for each language that some sequences of phonemes do not occur, e.g., in English morphemes a voiced member of the voiced-voiceless consonant pairs does not occur next to an unvoiced member: we have *dz* in *adze*, *st* in *west*; but we do not find *tz*, *ds*, or *zt*, *sd* (except in unphonetic spellings). We therefore define a new set of elements, as follows: given the phonemes $p_1, \ldots, p_n, p_{1'}, \ldots, p_{n'}$, if no sequence consists of both primed and unprimed phonemes, i.e., if $\exists p_i p_j \ldots p_m$ and $\exists p_{i'} p_{j'} \ldots p_{m'}$ but for every i, j between 1 and n, $\nexists p_i p_{j'}$, $\nexists p_{i'} p_j$, we define a phonemic component $'$ whose length is m units, and phonemic components $p_i, p_j, \ldots, p_m$ each of one unit length, and say that the combination of components $p_i p_j \ldots p_m$ represents the phoneme sequence $p_i p_j \ldots p_m$, and the component $'$ simultaneous with $p_i p_j \ldots p_m$ represents the phoneme sequence $p_{i'} p_{j'} \ldots p_{m'}$. Such a redefinition would not be worthwhile, unless the components have some relevant properties (e.g., $'$ here indicates voicelessness) and unless a distinguished set of phoneme sequences is involved (e.g., all sequences of certain phonemes). The result of the redefinition is that instead of the $2n$ phonemes with which we started, we have $n + 1$ phonemic components (one of them extending over sequences of the others), and the restriction as to combinations of primed and unprimed need not be stated since those combinations cannot be made with the new components. Thus we write *ædz* for "adze" and *we'zd* for "west"; the effect of the $'$ stretches here over the d as well as the z; the phonemes s, t, etc., no longer exist since they are simply combinations of $'$ with z, d, etc., and we no longer need a restriction that ds, etc., don't occur because ds, etc., cannot be formed from the new phonemic components. These newly defined phonemic components are free of the phonemic restriction of occurring only successively and of having no defined length. The new phonemic components (such as the $'$) can occur simultaneously with ordinary phonemes, and can have a length which is an integral multiple of the lengths of others. Each different combination of successive portions of one or more simultaneous phonemic components is equivalent to a phoneme. The new set of elements has more regular rules of occurrence than the original phonemes had; but the elements themselves are less simple.

Examples from morphology: Certain words which occur in unique neighborhoods, in which no other words occur, can be segmented into morphemes each of which has the natural neighborhoods of some particular morpheme class. Thus *who*, *which*, etc., are segmented into a connective *wh-* operating on two sentences, and a pronoun *-o*, *-ich*, etc., which occupies the position of the first noun in the second sentence (4.2.2.4). The cost is that we have here two small sets of morphemes (*wh-*; and *-o*, *-ich*, etc.) which can only occur together.

A somewhat similar method is used in reducing what is called grammatical agreement to regular sequences of morpheme classes. If a plural subject takes only a plural verb, we say that there is a single morpheme "plural" which occurs only on nouns but which has a discontinuous part that is attached to a following verb (before any permutation), if there is one. Then *The books are here* is simply the morphophonemic form of *The book 'plural' is here*. If the adjectives of a feminine noun receive a feminine suffix, we say that the suffix "feminine" occurs on nouns, but has a discontinuous part attached to the following adjective in the *N is A* form: *Le lion est las*; while *La lionne est lasse* is simply *Le lion 'fem.' est las*. (The other positions of feminine adjectives in respect to their nouns are due to later-operating transformations.) In this way we avoid the peculiarity of having a morpheme at one point (on verb or adjective) depend completely on a morpheme at another point (the preceding noun); the small cost is that some morphemes now have discontinuous phonemic composition (at least in certain neighborhoods).

It is not always possible to classify elements purely in terms of their degrees of freedom, without disturbing other considerations of classification—namely the sentence forms which they enter and the operations which they undergo. For example, we have ϕ_v, which introduces a new V (of subclass V_v) into its operand: e.g., $NtV\Omega \to NtV_v$ *Ving* Ω (*He is purchasing books*). And we have ϕ_s, which (in one of its subsets) introduces a new NV (of subclasses N_h and V_{-s}) into its operand: $NtV\Omega \to N_h tV_{-s} N$'s *Ving* Ω (*I regret his purchasing books*), where ϕ_z can operate if $N = N_h$: *I regret my purchasing books $\to$ I regret purchasing books*. We then note a set of operators (*begin, have*, etc.) which have the syntactic degrees of freedom of ϕ_v (i.e., the introduced V cannot have a new subject different from its operand) but the sentence forms of ϕ_s: *He began purchasing books, He began his purchasing books, He had a walk, He had his walk*, but $\nexists$ *He began my purchasing books*, $\nexists$ *He began my walk*. We can call these a subset of ϕ_s with reduced degrees of freedom; or a subset of ϕ_v on which an analogizing transformation (based on ϕ_s) can act: *He had a walk $\to$ He had his walk*.

6.2.   *Sets of complementary elements*

Another type of generalization is in the dependence of elements on neighborhood. Morphemes, for example, are initially defined as sequences of phonemes: *pulled* consists of two morphemes, phonemically *pul* and *d*, *pushed* has two, *puš* and *t*. Since these *d* and *t* morphemes are complementary as to environment (*d* only after morphemes ending in voiced sounds, *t* never there), we say, as with phonemes, that these are two positional variants of one past-tense morpheme; the latter is thus no longer restricted by the final sound of its predecessor. However there are various verbs to which the new morpheme cannot be added; when these verbs occur in environments where the past-tense morpheme occurs, we find either a changed verb (*take—took*) or no change (*I cut it now—I cut it yesterday*). We therefore change our definition of morphemes so that they are no longer sequences of phonemes but changes in the phonemic composition: usually additions of phoneme sequences, but now also zero addition (as past tense after *cut*), and in some cases replacement of one phoneme by another (e.g., after *take* the past-tense morpheme consists of the replacing of *ey* by *u*, yielding *took*). We now form a past-tense morpheme as a disjunction of all these environmentally complementary changes, and find that it indeed occurs after every verb of the language.

Whole classes of words or of operators can be eliminated. For example, the initial definition of a transformation, as a mapping between sentence sets which preserves the acceptability-inequalities within them, leads to a small set of adjunction-transformations which add quantifiers to nouns (*Men came ↔ Two men came*), adverbs of degree to adjectives and verbs (*He is youthful ↔ He is very youthful*; *I forgot ↔ I quite forgot*), etc. The quantifiers are similar to adjectives on the noun. But the adjectives are derived from a conjoined sentence:

Young men came. ↔ [*Men came*] *wh-* [*Men were young*],

whereas this transformational source is unavailable for *Two men came* because ∄ *Men were two*. However, there does exist *The men were two in number, The men numbered two*, etc., from which we can derive *two men*, just as from *The package was 2 lb. in weight, The package weighed 2 lb.*, we can derive *the 2 lb. package.*

As to the adverbs of degree, they are similar to other adverbs. But the other adverbs are derived from

He spoke deliberately ↔ *His speaking was deliberate,*

whereas this transformational source is unavailable for *I quite forgot* because ∄ *My forgetting was quite*. However, there are synonyms (6.5) of *quite* which do have the usual adverbial source, e.g.,

My forgetting was complete,

and we can say that in the transformation above this adverb receives optionally two phonemic shapes, both *I completely forgot* and *I quite forgot.*[1]

---

[1] The method of taking phonemically unrelated but environmentally complementary segments to be "suppletive" variants of one morpheme is well known in such cases as *is, am* (although here the change is not optional).

6.3.   Reducing one element to another

Much of the regularization of grammar consists in showing that some word or word subclass which had unusual restrictions as to combinability was the resultant of other entities (sometimes already recognized in the grammar), whose combined properties yielded precisely (or almost) the restrictions in question. A semantically obvious example of this situation is seen in paradigms: For example, in languages which have complicated paradigms, we can analyze the many suffixes of the verb into various combinations of person, tense, and number. These latter become the sets of the real morphemic elements of the language, instead of the morphemic elements being the actual suffixes which combine all these three together. Each of these three sets of elements has relations to particular other operators in their sentence: e.g., the plural suffixes on the verb, no matter what the person and tense of the suffix may be, occur with *and* or plural suffix on the subject noun. Such relations become more complicated to state if we take the more obvious whole suffixes of the verb as unit morphemes.

A special case of this is the reduction of the *wh-* connective to *and* plus a CS_{meta} which states sameness of individual for the repeated noun (5.7.2). When this is done, the choice of S_2 is determined simply by *and* and the double occurrence of a noun is required by the S_{meta} which asserts the sameness of individual. Both *and* and the S_{meta} in question exist in the language independently of *wh-*. The *wh-* is then simply the resultant of a ϕ_m which replaces ... *and* ... CS_{meta} by *wh-*. One advantage of this reduction is the following: It has been seen that *wh-* requires that S_2 be permuted so that the repeated noun is in first position (4.2.2.4). If *wh-* were a member of ϕ_c it would be the only case of an incremental operator requiring a particular paraphrastic operator, and the semigroup of incremental operators would be compromised. If *wh-* is a resultant of ϕ_m, then the requirement of a ϕ_p is not surprising, because various ϕ_m act only after particular operators; by no means all products of the paraphrastic operators occur (4.3.1).

The reduction of one conjuction to another has some similarity to the elimination of synonyms (6.5). However, for the most part the elimination of a conjunction is not a matter of finding some other conjunction whose synonym it is. Rather, we try to identify a conjunction C_a in $S_1 C_a S_2$ by what kind of conjoined sentence CS_3 is needed to ensure its acceptability: i.e., what kind of CS_3 assures the acceptability of $S_1 C_a S_2 CS_3$ for arbitrary S_1, S_2. If such a type of S_3 is found, it will state a C_a-type relation among the words of S_1 and S_2. If the S_3 simply contains nominalized $S_1 n$, $S_2 n$ then it is simply a transform of $S_1 C_a S_2$ and nothing was gained by using it to replace C_a; this is the case if we replace *She left because they phoned* by *They phoned and she left and their phoning caused her leaving*. But if the acceptability-ensuring S_3 relates the words of S_1 separately to those of S_2, in a regular manner, then it expresses non-conjunctionally the special meaning of C_a, and then all that is needed in the position of C_a is an *and* to express its conjunctional function. In the presence of this S_3, the C_a could then be replaced by *and*, since the substantive effect of C_a would be expressed by the S_3.

6.4. *Reconstruction of zeroed elements*

The most powerful regularizing operation is the zeroing of locally reconstructible material. There is, however, the danger of going beyond the stated conditions and making groundless reconstructions for the sake of regularization. For this reason, we summarize here the major kinds of regularization that are obtained from application of ϕ_z or of the automatic zeroing which is included in ϕ_m.

The automatic zeroing is the case of a morpheme getting a zero phonemic shape when it operates on particular operands. An example is the past-tense morpheme, which has variant forms on various verbs (*walked, took*, etc.) and on some verbs (ending in t) has the form zero: *He cut it yesterday*. Further operators operate on zero variants of morphemes just as they do on phonemically tangible variants. An example of this may be seen in the following case: If we compare (1) *He will wash it* with (2) *Will he wash it?* (and (1) *He can wash it*, with (2) *Can he wash it?*, etc.), we see that in forming the question (2) the tense-auxiliaries permute with the subject. If we now consider the corresponding (3) *He washed it, He washes it* with (4) *Did he wash it? Does he wash it?* we see that the apparently peculiar positions of the past- and present-tense suffixes can be explained without special transformations if we say that their initial position is like that of the auxiliaries (1) but, being suffixes and not pronounceable separately, they move to after the verb immediately following them (3). In the question (4), however, they have permuted with the subject and therefore no longer have a following verb to accept them; they are then pronounced on the word *do* which functions here purely as a phonetic carrier for them.[2] If we finally consider the parallel case of (5) *They wash it.* (6) *Do they wash it?*, we see that the presence of this superfluous *do* in the question-form (6) can only

---

[2] Not as verb, since the verb appears later on; not as auxiliary, since the corresponding assertion *He washed it* shows that there is no auxiliary present. This *do* is superfluous, and is not one of the morphemes of the sentence.

be explained as a phonetic carrier for the zero variant of the present tense which, like the *-ed, -es*, had moved (invisibly) to after the verb (5), but had permuted with the subject in (6), from its initial position before the verb. The zero variant is thus real enough to permute and require a phonetic carrier, which is visible in (6).

A somewhat similar evaluation can be made of other zero variants of morphemes, including those under conjunctions (*He came and spoke; He spoke and she too* $\xleftarrow{\phi_z}$ *He came and he spoke; He spoke and she spoke too.*), where we can say that it is only the phonemes that are zeroed: the affected morpheme continues to exist in the short forms also, but with zero phonemic composition. The mapping $S_i \rightarrow \phi_z S_i$ is however not isomorphic for some forms of S_i. For example, while there is no other source than the one above for *He came and spoke*, some zeroed forms can be derived from two different sources, as in:

I left him, he being ill at ease $\xrightarrow{\phi_z}$ *I left him ill at ease.*
I left him, I being ill at ease $\xrightarrow{\phi_z}$ *I left him ill at ease.*

Furthermore, for some S_i, $\phi_z S_i$ is the same sequence of word classes as $\phi_x S_j$ (for some x, j) so that further operations can be carried out by extension on the words of $\phi_z S_i$ according to the transformation-successions permitted for ϕ_x. This was seen in the analogic operations (4.2.4).

The zeroing which is necessary to obtain the rather obvious regularizations exemplified above can be readily extended to reconstruct, for any not fully regular sentence S_i, a source which is more regular than S_i, provided that the zeroed material can be determined, by standard linguistic methods, from what remains in S_i. This includes, for example, the replacement of adjectival occurrences of noun by *CS* formed with the aid of appropriate (or constant) words, e.g.,

The bookstore closed ← *The store closed. wh-*
   The store sells (or: *deals with*, etc.) *books.*
The book list is missing ← *The list is missing. wh-*
   The list lists (or: *names*, etc.) *books.*

Without assuming the zeroed appropriate words it would have been impossible to obtain a sentential source for the adjectival occurrence of *book*.

A different condition which can be regularized by assuming the zeroing of appropriate words is to be found in certain ϕ_s verbs, e.g., *expect* in $\phi_s S$: *I expect that he will fail, I expect him to fail*. These are the ones which occur in a restricted way with nouns (instead of sentences) as objects, as in the apparent *NVN*: *I expect him.* For all or certain of the noun objects of many of these verbs, one can show that $NVN \xleftarrow{\phi_z} \phi_s\, S$, where the zeroing is of appropriate words in the S: *I expect him* $\xleftarrow{\phi_z}$ *I expect him to come* (or: *arrive, be here*, etc). Thus *come* in S is an appropriate word for *expect* in ϕ_s; but *fail* is not, hence is not zeroed above.

Each set of appropriate words which can be zeroed is a set of local synonyms, of the type described in 6.5; and it is the local synonym set which is zeroed. Although this generalized zeroing does not have the strong restrictions and formal justification of the original zeroing of 4.2, it has an added justification in that the determinable material supplied for the source, and then zeroed, explains certain properties of the normally found form which is presumed to be derived by zeroing from that source: e.g., why ϕ_s like *expect* have N (instead of S) as object, or why certain $S_1 CS_2$ have no reduction of acceptability even though word repetition is not present.

6.5. Synonymless sentences

Just as the rules of zeroing enable us to fill out the existing sentences into regular types of elementary sentences, so they can be used to fill out existing SCS into regular types of elementary $SCS \ldots CS$. As we have seen, we can regularize SCS acceptance, and then word repetition within SCS, by assuming zeroed adjoined CS. This last method can readily be extended to derive all $SCS \ldots CS$ from a simpler source than the immediate needs of grammatical regularization might have suggested. We consider the fact that, for each sentence S_i which is understood by a receiver, the receiver must know a set of metalinguistic statements, $S_{meta(i)}$, which contain a grammatical decomposition or description of the sentence and definitions of its words. An original $S_i\, CS_{meta(i)}$ would be zeroed to S_i because the $S_{meta(i)}$ could be reconstructed by the receiver who understands S_i; hence $S_i\, CS_{meta(i)}$ is a source for S_i. We now define a relation of local synonymity: $\{X\}$ is the set of local synonyms of Y in a prime sentence p (i.e., in a particular kernel sentence or carrier) if each member of $\{X\}$ is a synonym of Y in p, although not necessarily a synonym of Y elsewhere. (Y is itself a member of $\{X\}$.) We can then assume that the source of S_i can contain various CS which state the local synonyms, antonyms, classifiers, etc., of the words of S_i as they are known to the receiver.[3] This will vary for

---

[3] It is a hypothesis of formal linguistics that the replacement of a word or word sequence W by a local synonym of W (including certain forms of definition of W) can be effected, for a given language or sublanguage, without affecting the environing sentences, except for a distinguishable factor of style, whereas replacement of W by anything else would affect the environment. Hence in principle synonymity can be determined from a corpus of experimentally obtained texts instead of from a dictionary or from the direct statements of a receiver.

different audiences; in particular, users of scientific articles know particular synonyms, classifiers, etc., of the specialized vocabulary of those articles. It is therefore possible to normalize sentences, especially in scientific articles, in the following way:

1. For each set of local synonyms $\{Y\}$, a representing member Y_r is chosen, and if a sentence contains any other member Y_i of $\{Y\}$, then Y_i is replaced by Y_r and the statement 'Y_r *is a local synonym of* Y_i' is added by C and then zeroed:[4]

$$S_{(Y_i)} \xrightarrow{\phi_m} S_{(Y_r)} CS_{syn(Y_r, Y_i)} \xrightarrow{\phi_z} S_{(Y_r)}$$

The synonymities may involve affixes. Thus if *to yellow, to pale,* are synonyms of *to become yellow, to become pale,* the zero verbalizing suffix is just a synonym of *become*; hence we do not need a verb *to pale* by the side of the adjective *pale*.[5]

2. If the S contains an antonym of Y, i.e., a word which is synonymous with *not Y* or *opposite of Y* or the like, we normalize correspondingly to $S_{(not\ Y_r)}$, etc. This incidentally gives a transformational basis for expressing that words which have what we may call "grammatically negative" properties contain *not* (e.g., *doubt* ← *not believe,* etc.).[6]

3. If S contains any word X_i that is a member or case of a vocabulary classifier $X_{cl(i)}$, we may add $S_{cl(i)}$ (i.e., a sentence specifying the classifier) if $X_{cl(i)}$ or some other member of it (or of contrasted classifiers) occurs elsewhere in the sentence or in the discourse neighborhood. E.g.

They described Vesuvius ← *They described Vesuvius, which is a volcano.*
His speaking was hesitant ← *His speaking was in a hesitant manner.*

This would be used if *volcano* or other volcanoes are mentioned; or if other adverbs of manner, or adverbs contrastedly not of manner, appear nearby. This makes it possible to relate words to other (neighboring) members of their vocabulary classifier. The vocabulary classifiers are the last N in sentences of the form X *is a* (*member of,* or: *case of,* etc.) N. These classifiers are not merely semantic; they distinguish refined subclasses of their subject X which have distinct properties in the grammar of the language or of an appropriate sublanguage: e.g., volcanoes as against

---

[4] This activity is a case of ϕ_m, producing an alternant form to $S_{(Y_i)}$.

[5] The representing synonym can eliminate various restricted words: e.g., *credible* would be represented by *believable*. And if Y is allowed to range over word sequences, the representing synonym can eliminate idiomatic phrases.

[6] Similar methods could be used to replace a word by a synonymous combination of a simpler (or a classifier) word plus a modifier, as can be done in many dictionary definitions.

mountains, and both as against mammals; adverbs of manner as against adverbs of time; quantities on the scale of weight (*2 lb. is a weight*) as against the scale of length (*3 feet is a height*); etc.

4. If S contains a classifier (or pro-word) X_{cl} without a specified member of it, then we define an operation of a ϕ_z type which replaces X_{cl} by the appropriate member if the latter is transformationally identifiable, or otherwise by a disjunction or conjunction of the members of X_{cl}, always adding that these are members of X_{cl}. E.g.,

A large salmon leaped up. The fish missed the rock and fell back. ←
... The salmon, which is a fish, missed... ←
...The salmon missed...and a salmon is a fish.
A large carnivore escaped. ← *X or Y...or Z escaped.*
X and Y...and Z are large carnivores.[7]

All these methods, if used to their fullest extent, transform the set of kernel sentences K into a set of $RCD...CD$ sentences, i.e., a set R, of sentences with reduced vocabulary, to each of which certain dictionarylike sentences D are adjoined. The sentences of R contain only a representing member (preferably an unambiguous one) of each local synonym set;[8] they do not contain such derivable words as the " grammatically negative " verbs; they do not contain classifier words except in the last position of X *is a* X_{cl} sentences. The vocabulary of the R sentences consists of the symbols for synonym sets Y or their unambiguous representing members, and contains no synonyms. The grammar and vocabulary of the source sentences R and D are therefore simpler and more orderly, and more amenable to formalization. If the representing synonyms and the forms of dictionary sentences are chosen in fixed ways, the set K is isomorphic (under all remaining transformations) to the set $RCD...CD$, and maps homomorphically onto R; but since the sentences D are known and zeroable, the information in each R_i (and $R_i CD_{i1}...CD_{in}$) is the same as in the corresponding K_i.

---

[7] In discourse analysis (5.8) we regularize sentences not necessarily toward a standard elementary and concrete form, as here, but rather toward maximum similarity with each other. For this purpose, some of the regularizing transformations may be taken in the opposite direction from the one given here. E.g., we might regularize

They described Vesuvius → *They described the volcano* (*which is called*) *Vesuvius.*

[8] If a local synonym set Y contains a word Y_r which occurs in no other local synonym set, Y_r can be chosen as an unambiguous representing member of Y. If, failing this, Y contains a pair of members Y_s, Y_t (Y_t could be a classifier of Y), such that no other synonym set contains both, then a new hyphenated word Y_s-Y_t can be constructed as an unambiguous representing member of Y.

The fact that in the source sentences $RCD \ldots CD$ each word is accompanied by its classifiers, if any, makes various transformational and discourse operations on these sentences much easier to formulate. The synonym-eliminating operation described here does for arbitrary sets of words and word sequences (including idioms), which have identical range of neighborhoods, what the paraphrase-eliminating operation (5.2) did for neighborhood-preserving changes on grammatically recognized sets of words.

Fixed dictionary sentences of the type D cannot be constructed except in tightly organized portions of language, such as the sentences of a particular science. A simpler case arises in analyzing a discourse, where we can increase the similarities among residual sentences by using such D sentences as would reduce two different discourse sentences to the same R sentence (the D sentences being drawn from among those which are common knowledge to the participants, and hence recoverable and zeroable). In any case, the methods of 6.5 cannot be used fully over a whole natural language.

6.6. Infrasentences

Throughout this chapter we have been considering situations in which the methods that established useful relations among sentence elements (relations useful for characterizing sentences) can be extended to provide even more regular relations among the elements, but at the cost of going beyond the more compact apparatus. It is never a question here of formulating a more general structure, of which language is a particular case. We are dealing here only with structures that characterize precisely language. However, the compact structures leave unstated certain relations of certain particular elements: e.g., the restriction between z and d in 6.1, the dependence of plural verb on plural noun in 6.2. On the other hand, the generalized structure leaves unused some of its capacity: the definition of phonemic components in 6.1 would make it possible to describe a language in which all phonemes were composed of simultaneous components of different lengths; the method of 6.2 would suffice to describe a language in which all morphemes had discontinuous portions variously located in the kernel sentences. This problem is unavoidable: because, in view of the fact that language is an open and always-developing system, sentences cannot be described by any completely utilized structure of elements and operations on them. A goal of fully independent elements using up all the well-formed sequences is unreachable. Either the structure is fully utilized but some further relations remain undescribed or else the further relations are shown to be generalizations of the original structure, but these generalizations are not fully utilized. However, it should be clear that the generalizations discussed here are those sufficient precisely to include what would otherwise be the exceptions or the limited-domain relations in sentence segments; if these further relations were used over the whole domain on which they are defined, we would still have a system which is only like language (perhaps a closed sublanguage) and not like anything else.

The element definitions of 6.1 and 6.2 provided for exceptional relations among elements. The zeroing of 6.4 and the synonym elimination of 6.5 provided a regular transformational source for a more narrowly defined set of residual sentences. Now, however, we must consider a more extreme case, one in which the transformational method, based initially on a relation among sentences, leads necessarily beyond sentences. This situation arises because, once we have determined the transformations which are necessary for the sentences of a language, we find that the domains of the transformations and the set of transformational products (i.e., the rules for their successively operating one on another) are not fully regular: again, there are exceptions, necessarily. First, we find how the exceptions can be seen as extensions of the existing domains and successions. Then we include these extensions in the definitions of the transformations. When we do so, we obtain not new sentences (which we would not want to obtain) but transformational operands or resultants which are not sentences but which become sentences upon further regular transformations. In short, the transformational characterization of sentences can be regularized not by extending the set of sentences (which need not be done by any method in this chapter) but by generalizing the domains and ranges of some transformations to being not only sentences but objects, called infrasentences, which become sentences by further existing (previously established) transformations.[9]

In linguistics, the situation is well-known from ordinary morphophonemics. Given both ∃ *I can not go* and ∃ *I can't go*, we form *I can not go* in the regular grammar, and then add an operation which changes *not* to *t* in stated environments. But given *The knife fell*, *The knives fell* we cannot obtain the latter without first forming the non-existent ∄ *The knife pl. fell*, in which alone the morphophonemic change *nayf* → *nayv* acts. Similar situations of non-existent intermediate forms arise elsewhere.

The search for independent elements had already led at certain points to infrasentence structure. This happened for certain automatically occurring

[9] Naturally, we would not define special new transformations just for the purpose of making these objects into sentences, for then the whole procedure would be *ad hoc*.

elements. E.g., certain nouns do not appear in a sentence without a quantifier or pro-adjective: e.g., *A hat fell. The hat fell. Hats fell.*; but *∄ Hat fell.*[10] We could construct a source *Hat fell* with automatic *a*, but then we would no longer have a sentence as source. Somewhat differently, we have a set of kernel sentences whose sole verb is *be*:[11] *N is A* (*Butter is yellow.*), *N is PN* (*Butter is on the table.*), *N is D_{loc}* (*The butter is here.*), and the various *X is N_{cl}*. All these could be derived from a simpler (nonsentence) form *NA*, etc., in which automatic *be* is then inserted.

Infrasentence sources are also reached in the operands of certain transformations. Certain ϕ_s and ϕ_c operate on sentences which contain no tense. Thus in *I know that he is there* the ϕ_s operates on *He is there*; in *I know that he was there* it operates on *He was there*; in *I know that he should be there* it operates on *He should be there*. But in *I ask that he be there* and *I ask that he should be there* (which can be considered as only morphophonemically different), the ϕ_s operates on an infrasentence *He be there* with added *should* as variant.

The major source of infrasentences arises when we seek to regularize the operands for particular operators. Thus, given that *What did he take?* is derivable as a transform of its answers *He took a book* or *He took a pen*, etc., we would like *What did he do?* to be similarly derivable as a transform of its answers *He smoked* or *He wrote a letter*, etc. This can only be done (as in 4.2.4) by:

He smoked. $\xrightarrow{\phi_v}$ (1) *He did smoking.* $\xrightarrow{\phi_s\phi_c}$ (2) *I ask whether he did smoking or writing a letter ... or Ving.* $\xrightarrow{\phi_z}$ (3) *I ask what he did.* $\xrightarrow{\phi_m}$ (4) *I ask: what did he do?* $\xrightarrow{\phi_z}$ (5) *What did he do?*

The *did* of (1–3) and *do* of (4–5) is the verb operator brought in by ϕ_v. The *di-* of (4–5) is the phonetic carrier of the tense when it permutes with the subject (see 6.4). This derivation is identical with that of all other questions, except that it involves marginal sentences: the member *do* of ϕ_v does not operate upon *smoke, write a letter*, and many verbs, although it operates on certain ("occupational") verbs, as in *He does bookbinding*, and operates perhaps in all verbs when they have quantifiers: *He did some smoking, He did a lot of smoking*. We can fit this situation into our description by saying that for some *K* and some ϕ it is only products >1 of the ϕ that can operate: e.g., here only the product $\phi_z\phi_s\phi_c\phi_v$ operates, and not any part of it separately. Or we can say that the resultants and operands of

---

[10] In *Hat-wearing is widespread*, the *hat* occurs in adjective position.

[11] All other verbs in this kernel set, e.g., *seems*, can be derived transformationally from *be*: they do not affect the acceptability-ordering of the pairs *NA*, etc.

certain ϕ, or parts of products, may be marginal sentences rather than acceptable sentences; only the final resultant must be an acceptable sentence. That is, certain $\phi \ldots \phi K$ produce sentences only when operated on by further ϕ.

The problem becomes sharper when it is the ultimate source that is not an acceptable sentence. Thus the synonymous *He threw a party. He gave a party. He made a party. He had a party* have the properties of $\phi_v K$ or $\phi_s K$, as in *He gave a look* (even *He threw a look*) from *He looked* or *He threw a glance. He gave a glance* from *He glanced*. But $\nexists$ *He partied* (except in recent colloquialism). ϕ_v would be more regular if it could operate on such infra-K as well as on K. The infra-K produce sentences when operated on by ϕ.

A more interesting case is that of the *NVNPN* and similar kernel sentences: *He gave a book to the boy. He removed a page from the book. He attributed the plan to her. We left it aside.* With a certain amount of special morphophonemics, we can derive these by ϕ_s from *NVN* kernel sentences:

He caused the boy to have a book. He caused the book to lack a page.
He claimed that she made the plan. We caused it to be aside.

In some cases the new kernel sentences may be quite marginal (e.g., *It is aside.*). In any case such transformations are worth formulating only if there is a regular way of replacing the existing $V \ldots P$ by the $\phi_s \ldots V$ of the proposed source.

When all such methods have been utilized, we obtain a set of kernel sentence and infrasentence forms, the latter transforming into sentences by ϕ_m or other ϕ. Aside from details of formulation, these are:

NV
NA
ND_{loc}
XN_{cl}
$NVN, NVPN$
NPN: *The tree is near the brook.*
$NN_{rel}N$: *John is the son of Jones.*[12]

These become sentence forms only when the class t of tenses, and also, for infrasentences lacking V, the word *be*, is inserted. In addition, for certain words as values of V, etc., these forms produce sentences only after they are operated on by certain ϕ (e.g., in *He partied.*). The forms above apply

---

[12] The N_{rel} includes all relational nouns, including, in the D sentences, *is synonym of*, etc.

not only to the language as a whole, but also to the reduced-vocabulary (R) and dictionary (D) and metalinguistic (S_{meta}) subsets of elementary sentences.[13] The classes of transformations necessary to derive all sentences from these forms are the same as those in the original list above; but some classes of transformations, particularly the ϕ_s, ϕ_z, and ϕ_m require additional members that operate particularly on the simplified elementary sentences indicated here and on the $SCS \ldots CS$ which have been built up metalinguistically (including by means of dictionary sentences).

As to the domain and range of transformations, these are now no longer subsets (proper or not) of the set of sentences $\{S\}$ but subsets of a set $\{Z\}$ related to $\{S\}$ as follows:

$$(z)z \in \{Z\} \quad \text{iff} \quad (\exists \phi)\phi z \in \{S\}.$$

Transformations are regular if they are defined on objects z such that there exists some transformation ϕ among the otherwise extant transformations of the language, for which ϕz is a sentence of the language.

If, then, we imbed the set of sentences and marginal sentences in a set that consists of them and also of such infrasentences as are required here, we obtain a simpler set of elementary forms and more regular operation of the transformations defined on them.

We never achieve a description in which all combinations of our elements occur: a language in which this were possible could carry virtually no information. But if in a given language we define elements whose combinability is least restricted, we will obtain the greatest correlation between differences of sentence structure and differences of information carried.

6.7. *From exceptions to extensions*

Because of the great complexity of linguistic data, any analysis leaves a residue of apparently aberrant elements or types of structures which are not overtly similar to the main ones, and of occasional individual elements or sequences which fail to satisfy some of the properties of their class. We try to show that these are the results of particular combinations of general rules of the grammar (e.g., particular products of transformations). Where this proves impossible, we may have to change our definitions, giving the elements more degrees of freedom, or accepting a weaker structure of which the bulk of the language, the more regular part, is a special case.

Each such generalization makes room not only for extant linguistic material, but also for material which differs from the main material of the language; and it does so at the cost of destroying some property, often

---

[13] In the R infrasentences, words which appear synonymously in more that one of the above types are located primitively in only one. Thus *pale* (6.5:1) would be a value of *NA* only and not of *NV*.

some correlations and interpretations, which obtained for the tighter structure. Thus if we change the partial transformations into transformations over the whole set of sentences, we lose the uniqueness of decomposition for a graded sentence, and also the distinction between the transformations which really apply to the whole set of sentences and those that are really partial.

The stock of idioms and similar individual exceptions in language is such that, after all such generalization is carried out, there always remain many idiomatic exceptions. For these, the only hope is to formulate the general rules in such a way that the individual exceptions can be stated as extensions of the domain of one rule or another, beyond the boundaries allowed in the general rule (2.6, 4.2.4).

6.8. Toward reduction of grammar

The methods of 5.6, 7 and Chapter 6 regularized parts of grammar, by replacing various restricted combinations of elements by unrestricted ones. If such methods are carried out as far as they can go, they have the effect of eliminating a considerable part of the grammar of a language, and of replacing it by a network of morphophonemic operations. The fact that certain grammatical restrictions have been replaced by morphophonemic ones, which are necessarily paraphrastic, means that these grammatical restrictions did not really contribute an increment of meaning to their sentences. Thus, such reductions do not affect the meaning-bearing properties of the language. But the fact that the grammatical restriction is removed shows, in some situations, that what appeared grammatically as a restriction on sentence-relations did not in fact exist, and that a wider system of sentence-relations is possible in the language than the traditional grammar had shown.

Two results may be surprising: that one can remove parts of the grammar without affecting the language, and also that one can seem to extend the language with no more than a morphophonemic change.

We consider the first: In a language which has a dual in addition to a singular and a plural, all forms N *dual* and *two N dual* can be obtained by a morphophonemic operator from *two N*.

The derivation would be, e.g., for N_1 dual t V_1:

(1, by ϕ_c) N_1 t V_1 *and* N_1 t V_1 (for counted N_1)
(2, by $\phi_p \phi_z$) $\rightarrow N_1$ *and* N_1 t V_1
(3, by ϕ_m) $\rightarrow$ *two* N_1 *dual* t V_1

Here the *two* is zeroable, yielding

(4, by ϕ_z) $\rightarrow N_1$ *dual* t V_1

If the ϕ_m only produces *two* out of *and* N_1, but does not add a dual affix, the resultant would be (3′) *two* N_1 *t* V_1. Thus (3) with the dual is only morphophonemically, paraphrastically, different from (3′); however, in the absence of this superfluous (pleonastic) dual affix, the *two* is no longer zeroable in (3′) as it was in (3) *two N dual*. The same applies to languages which have plural, as in English. The plural, *nN pl.*, with some number $n > 1$ or with a pronoun for indefinite such *n*, is a morphophonemic form of *N and N...and N* with *n* occurrence of *N*. As above, if the pleonastic suffix *pl.* is omitted from the morphophonemic operator, we obtain the form *nN*, which is equivalent to *nN pl.* except that in the absence of *pl.* an indefinite *n* can no longer be zeroed.

The second result above is more complicated: If we consider languages that have tense-affixes, such as English, we find that each language has particular tenses, e.g., for particular kinds of past, and verbs seem to be restricted to having only these tenses. However, there is no restriction as to what ϕ_s of time may appear with each verb, and the tense-affixes can be taken as simply a morphophonemic change which depends on what time-ϕ_s the verb has received. Roughly, if the verb receives a time-ϕ_s such as *yesterday, in the past*, it also receives a past tense affix; and so for the other tenses. The time-affixes are thus classifiers of particular ϕ_s of time, and their one contribution is that only in their presence can we zero the indefinite time-ϕ_s which is within their range (e.g., in the presence of *-ed* we can zero *in the past* or the like). In spite of the importance of tense in traditional grammars, it thus appears that verbs in a particular language are not restricted as to time; and the classifying of time-ϕ_s into particular tense-affixes does not constitute any restrictive structuring of the set of sentences, but only a selective calling of our attention to a particular time division. However, the paradigmaticity of time, i.e., the fact that in every sentence (and certain clauses) the main verb must have one of the tense-affixes, means that every sentence must have had a time-ϕ_s (which may then have been zeroed); this leads to artificial meanings for the tense-affixes in cases where the time-ϕ_s is irrelevant (as in the narrative present).

A more striking example is to be found in the comparative. Overtly, English seems to have a comparative conjunction: (1) *The first solution is darker than the second* (*is*) is apparently a conjunction C_c on *The first solution is dark, The second is dark*. This C_c has a disturbing restriction on repetition: C_c can connect SC_cS not to another S but only to another SC_cS: $\exists C_c(S_1, S_2)$ as in (1), and $\exists C_c(C_c(S_1, S_2), C_c(S_3, S_4))$ as in (2) *The first solution is darker than the second* (*is*) *more than the third is darker than the fourth*; but $\not\exists C_c(C_c(S_1, S_2), S_3)$ as in (3) $\not\exists$ *The first solution is darker than the second more than the third is red*. In contrast, other C

repeat without restriction: $\exists\, C(S_1, S_2)$ and $\exists\, C(S_1, C(S_2, S_3))$ and $\exists\, C(C(S_1, S_2), S_3)$ as in *I left because it was late because I had to be up early*. Now, the attempt to derive the comparative forms from simpler English material shows that they can all be obtained, in a regular way, from *n exceeds m by p* and *n equals m*, where *n, m, p* are numbers: (1) ← (4) *The amount of the first solution's darkness (on some scale) exceeds (by some amount) the amount of the second solution's darkness*. The form that is called comparative, (1) *The first solution is darker than the second*, is obtained from this by a morphophonemic operator, which also yields (2) above from (5) *The amount of excess of the first solution's darkness over the second's exceeds the amount of excess of the third's over the fourth's*. But it then transpires that this form is not restricted to repeating on itself, i.e., on another pair of compared sentences, but can repeat on a single new sentence: e.g., (6) *The amount of the excess (on some scale), of the first solution's darkness over the second's exceeds the amount of the third solution's redness (on some scale)*. The set of source sentences is therefore not restricted, and such restriction as appears in the grammar holds only for the morphophonemic operator which operates on (4) and (5) but not on (6), and hence produces (1) and (2) but not (3). Since all paraphrastic operators have restricted operands, the restriction on this morphophonemic operator is by no means as disturbing as a comparable restriction on a particular conjunction C_c.

The considerations above make it clear that not everything in grammar determines what can be said in a language, though it may affect (in ways that are so far not investigable) the framework to which the language users relate what is said. This helps explain the fact that while not everything is allowed by the grammar of a particular language, it is nevertheless possible in many cases to translate into one language sentences which come from a differently-structured other language.

The grammar-reducing methods discussed here have not yet as yet been carried out to their fullest extent, and we do not as yet have a picture of where they stop and what is irreducible in grammar. It is clear that some grammatical restrictions are substantive, and can not be eliminated by making them morphophonemic. This applies to many of the operations in the restricted inner set ϕ-ψ (4.3.1). For example, the adverbs of manner are a subset of ϕ_s which are restricted in English to S_n form 5 which is not available for the verb *be* (4.2.2.3) *His selecting of books is slow.*; hence they do not occur on *be*-sentences and in particular do not repeat on other adverbs of manner: $\nexists$ *His being of large is slow*, $\nexists$ *He is large slowly.* $\nexists$ *His speaking's being slow is quiet*: Now, we could form a classifier *manner* for these adverbs, and then use the ϕ_s: *occur in a manner* (which would take *Sn* 4 rather than *Sn* 5) instead of the manner-adverb ϕ_s which require *Sn* 5: e.g., *His selecting books occurs in a manner which is slow*. We could then form *His being large occurs in a manner which is slow*. In this way we would actually be forming sentences which were impossible before, because *slow* could not be said before of *He is large*. However, it is doubtful that the existing grammar permits *occur* to operate on *He is large* or permits *in a manner* to operate on *occur*, so that it may be that *His being large occurred in a manner* is no more formable in the grammar than is *His being of large is slow*. In any case, whenever the elimination of a grammatical restriction leads to the formation of actual new sentences in a language, we have to consider whether a meaning can be assigned to these new sentences. Thus, those regularizations of grammar which result in extensions of the language must be decided with an eye to the intended interpretation.

7

The abstract system

7.1. Mathematical formulation of language structure

From the work of the preceding chapters, it is possible to formulate an abstract system, i.e., one whose objects are defined purely by the relations in that system, which is adequate precisely for natural language. The details are based on English and a few other languages; modifications would presumably be needed to account for still other languages. However, saying that this system is adequate for language means that no special properties or individual facts should be encountered in language except those that result from the formulation of the following types of operators and their domains.

7.1.1. For sentence forms

For sentence forms S, an axiomatic theory of string analysis is available (3.5), using: a finite alphabet of symbols (intended to represent word classes or morpheme classes); a (finite) set of finite sequences ("strings") of these symbols, the strings being collected into subsets such that A and B are members of the same subset iff

(C) $[(\exists S)$ C is part of S.   and. A adjoins C at p: iff:
$(\exists S)$ C is part of S.   and. B adjoins C at p],

where C is a string, p a stated point next to one of its symbols; and a rule of derivation which says that the result of adjoining a string of subset P to a string of subset Q is a string of subset Q. Those strings which are adjoined to no string are the elementary sentence forms of the language.

A recurrent dependence process (3.6) can tell, given $w_1 w_2 \ldots w_n$ (where w_i is the ith symbol of a sequence in this alphabet), whether the $n + 1^{th}$ symbol is a possible next symbol within a sentence form, or the beginning of a new sentence form, or neither. And in a quasigroup of words in a different alphabet, of symbols expressing the string relations of word classes, we can show that all sentence forms of the language are represented by the null reduced word (in the cycling cancellation automaton of 3.7). In all these cases the denumerably infinite set of objects which one obtains represents the sentence forms of the language.

These sentence forms yield sentences (of various nonzero degrees of acceptability) when their primitive symbols are replaced by words (or morphemes) which are in the domain of these symbols, the words being in turn obtained by a recurrent dependence process (3.2) on the phonemes established by the pair test (3.1) out of unanalyzed occurrences of speech.

7.1.2. For sentences

The abstract system adequate to characterize the actual sentences of natural language is an ordered 6-tuple $<\overline{N}, f^0, f^1, f^2, f^3, f^4>$.

We are given an alphabet of primitive arguments and function symbols. The arguments (written $\overline{N}$ in 7.1.2) are of two kinds: a finite set N and an infinite set 'q.' The function symbols are of four, possibly five, kinds: f^0, f^1, f^2, f^3, f^4. Each f is a finite set of operators; each operator introduces material which is concatenated to its operand, or introduces changes in its operand.

f^0 is a finite set composed of various subsets of operators whose domain is $\overline{N}, f^1$, or f^2. For each subset of each of these (N_i in N, etc.), f^0 does not change the class and subclass of its operand. Thus

$$(x)x \in N_i \supset f^0 x \in N_i,$$

and so for f_i^1, f_i^2. f^0 either changes the form of x or concatenates a member of f^0 to x.

f^1 is a finite set composed of various subsets of operators whose domains are in the primitive arguments. Some f^1 have one-place arguments, some have two-place arguments; in some languages there are f^1 with arguments of a larger, finite, number of places. They are defined (taking a one-place argument as an example):

$$(x)[x \in f^1(\overline{N}) \quad iff \quad (j, k)[x\overline{N}_j \geq x\overline{N}_k$$
$$or \quad x\overline{N}_k > x\overline{N}_j]].$$

Thus, each member x of f^1 concatenates x with the operand of f^1, and imposes a linear ordering on subsets of resultants. The two-place and other subsets of f^1 have corresponding definitions.

f^2 is a finite set composed of various subsets of operators, the domain of each subset of operators being all or particular f^1 subsets and all or particular f^2 subsets. They satisfy the following condition:

$$(x)[x \in f^2 \quad iff \quad (i, j, k)] f_i^1 \overline{N}_j > f_i^1 \overline{N}_k \cdot \supset x f_i^1 \overline{N}_j > x f_i^1 \overline{N}_k$$
$$.and. \quad f_i^1 \overline{N}_j = f_i^1 \overline{N}_k \cdot \supset x f_i^1 \overline{N}_j = x f_i^1 \overline{N}_k]].$$

Thus every member x of f^2 concatenates x with the operand of f^2, and preserves in the resultant the ordering of the operands. Corresponding definitions would be needed for the other operands of f^2, namely the other subsets of f^1 and also (since f^2 iterates) the concatenations, to any finite n, of $f_1^2 \ldots f_n^2 f^1$. However, a subset of f^2 (representing the ϕ which are not in ψ, 4.3.1) do not concatenate with all f^2 or with all f^1. The f^2 which represent ψ and are not restricted will be called $f^{2\psi}$. There are also subsets of f^2 whose domain is the set of pairs formed by two occurrences of f^1, each f^1 on its own operand and each f^1 with possibly various f^2 operating on it. In the case of f^2 operating on a pair of f^1, we require the resultants to preserve the ordering of the minimal acceptability (5.6.2) of the operand pairs. In a restricted way, specified in 7.1.2.2, f^4 appears in the operand of certain f^2.

f^3 is a finite set of operators defined on concatenations, to any n, $n \neq 0$, of $f_1^2 \ldots f_n^2 f^1$, and on f^3, and on f^4, but not directly on f^1. Like f^2, it preserves the ordering of its operands. It introduces certain changes into its operand, rather than concatenating material to it. The fact that f^3 changes its operand rather than adding to it limits the ability of an f^3 to repeat, because the argument of f^3 becomes unavailable after the f^3 has operated. This holds not only in obvious cases such as zeroing but also in permutation, which requires a particular D or CS.

f^4 is a small finite set of operators defined on f^1 and on f^2. It concatenates $x \in f^4$ to its operand, and its resultant represents a sentence S of the language.

This abstract system becomes a characterization of the sentences[1] of a language when:

(1) The members of N, f^1, f^2, f^4 and the incremental members of f^0, are morphemic segments, i.e., subsequences, determined by the recurrent dependence process of 3.2, of the phonemes established by the pair test (3.1) out of unanalyzed occurrences of speech; and the members of 'q' are any sounds (or written representations of them), including any sound sequences produced by the above system as occurrences of the language.

(2) The ordering of the resultants of each f^1 is the ordering of their normal acceptability as (elementary) sentences, or of the class of discourse or neighborhood in which they are normally acceptable.

When these conditions are satisfied, the resultants of $f^4 f^1$ are elementary sentences (i.e., f^1: arguments composed of N or 'q' (with possibly f^0 on them) → objects which, when operated on by f^4, are elementary members of the set of S); and f^2, f^3: $S' \to S'$ while the ϕ_c subset of f^2: $S', S' \to S'$;

---

[1] More exactly of the propositions, i.e., unambiguous grammatical meanings of sentences. A grammatically ambiguous word sequence appears in two or more different characterizations in this system.

(S' here indicates either S, or an S lacking only f^4). If we use the regularizations of 6.6, then the resultants of $f^4 f^1$ are elementary members of Z,[2] where Z includes S, and

$$(x)[x \in Z \quad iff \quad (\exists \phi)\phi \in f^2 \cup f^3 \text{ and } \phi x \in S].$$

That is, Z is the set of objects transformable into sentences. The required ϕ is in many cases an automatic ϕ_m as in the be of 6.6. Thus f^1 produces the set of structurally axiomatic S' from which f^2, f^3 derive the rest of the set of S'. (All sentences obtained without 6.6 are in Z, since every sentence participates in some transformation, 4.3.2.)

From the definitions, it follows that the f^1 are not iterable, since f^1 does not operate on f^1; hence the elementary sentences produced by the finite number of f^1 are themselves finite in number except for f^1 'q'. The f^2 (or some of them) are iterable without bound, so that although every sentence is finitely long, the set of sentences produced by the f^2 is denumerably infinite. The f^3 are also iterable, but since they have stringent conditions on their operand, the possibilities of their acting on a given sentence (unless further f^2 are brought in) soon fall off. One result of this is that if question arises as to whether a given part of a sentence is due to an f^1 or an f^2 (since some words are members of both), the decision can rest on whether or not iterating the given part in that sentence produces another sentence.

The replacing of K by f^1 and of incremental ϕ by f^0, f^2 is interrelated. The f^2 differ from the incremental ϕ in an important respect. Each ϕ consisted of all the changes made in an operand S to obtain a resultant S. Each f^2 consists of the unique part of the corresponding ϕ, which turns out to be the predicational part. Hence many f^2 operate not only on an S but on the S plus an N or the like. For example, $\phi_s(S)$: *He knows that* on S, is replaced by $f^2(N, S)$, as in *know* on (He, S). The f^2 are then structurally and semantically alignable with the f^1, which are predications operating on N and which, with their operands, replace K.

7.1.2.1. *Products of the operators*

The fact that the operators (except f^3) concatenate with their operands makes it easy to look upon the products (i.e., successive applications) of operators (including f^3) as words in the f alphabet. The f can be defined in such a way that each proposition (graded, i.e., unambiguous, sentence)

---

[2] In this case the set f^1 can be simpler, since the n-place members of f^1 (below) may be obtained by existing f^2 from two-place members of f^1 (e.g., *He threw the ball onto the field toward the right* $\in S$ might be obtained from *He threw the ball so that the ball was on the field and the ball was toward the right*, which is perhaps in Z rather than in S).

of the language has a unique characterization as a partially ordered structure (or ordered product, i.e., word, with certain portions commutative) of these operators and the ultimate operands (N, 'q'). The set of products of $f^{2\psi}$ is a semigroup, and each ordered product characterizes a sentence.[3] The set is a monoid if we define an identity in f^2 : $1 \cdot f_i^2 = f_i^2$. This disturbs the unique characterization of sentences, but in an unimportant way: each sentence still has a unique characterization except for insertions of the identity in the product.

If we take the whole of f^2, or the union of f^2 and f^3, the products which characterize sentences do not form a semigroup; f^3 operate only on operands of particular form. We can obtain a semigroup by defining $f_i^3 f_j = f_j$ in case f_j does not satisfy the conditions for being an operand of f_i^3. This destroys the unique characterization of propositions in respect to their f^3 (since each sentence could be said to have each f^3 in any position of its decomposition, the f^3 taking effect only if the conditions for it exist); but it does not destroy the f^2 composition of sentences, nor the semantic interpretation (since f^3 are paraphrastic).

If we further include the analogic transformations (4.2.4), f_{anal}, in the set $f^2 \cup f^3$, the sentence-characterizing products again are not a semigroup, since the f_{anal} have restricted operands. And in this case, even though f_{anal} are also paraphrastic, it is undesirable to define $f_{anal} f_i = f_i$ in case f_{anal} doesn't operate on f_i, because it does not seem much in point to say, for example, that each sentence may contain the passive at all points of its decomposition where a passive could not operate. However, it has been seen that the analogic transformations are not arbitrary changes but can be analyzed as products of f^2, f^3, and their inverses, with the special condition that for the inverse or for some other f in the product the domain is not a subset of the range of its operand, though similar to it. This mismatch, of the operand of the $n + 1^{th}$ f in a product and the resultant of the nth, produces a novel kind of sentence. We can therefore define inverses of f^2, f^3, and form the group of products of f^2, f^3 and their inverses (with identity included as above). We now form the set of reduced words by cancelling all pairs consisting of f_i^{-1} operating on f_i. This eliminates all the nonunique representations of a sentence which would arise from those f-products that contain some f_i which is then

---

[3] In addition, we have to consider the members of f^2 which are not in $f^{2\psi}$. These have restricted domain, e.g., *take* in ϕ_v which operates on *walk* but not on *talk* (*He took a walk; $\nexists$ He took a talk*). Many of these restricted members have such relations to other members as to make it possible to describe them as restricted morphophonemic variants of unrestricted members or disjunctions of members (e.g., *take* as variant of *have* in *He had a walk, He had a talk*).

immediately removed. The set of reduced words does not characterize the set of sentences, but has a certain relation to that set: It includes representations of all sentences, plus representations for all the possible analogically formed sentences that could at all be formed in the language (not only for minor mismatches as in the existing analogic forms but also for arbitrary and major mismatches), plus representations for negative sentences and negative analogic sentences (these being objects whose concatenation with their positive counterpart yields zero).

For sentences which are characterized by positive f, without inverses, the order of the f is not in question. However, once we admit inverses of f it becomes possible to represent a single proposition by more than one different product of f. The representation of each sentence also depends, naturally, on the particular f which are defined for a language; there may be alternative possibilities. However, even if we consider alternative sets of f that may be definable for a language, and alternative f products that may represent an analogic sentence, one thing remains true: every sentence is the result of some set of elementary differences from other sentences. That is to say, an object is a sentence if it is a member of one or more sentence pairs, the differences between the members of a pair being some sum of differences which appear in other sentence pairs.

7.1.2.2.   The operators for English

In English, and in many other languages, the set N is what might be called simple nouns.

In the various f^i, the main subsets (distinguished by their operands) follow; we omit the concatenation sign in the resultants.

In f^0: a few word modifiers ϕ_a such as *some* and perhaps *very*; some of these are not iterable although the undetailed description given above for f^0 allowed iteration.

In the set f^1:

$f^1 : \overline{N} \to \overline{N} f^1$
    (*exists*[4] in *John exists.*
    man in *John is a man.*
    ill in *John is ill.*)

---

[4] The words in N, f^0, f^1, and f^2 differ from one set to the other although the morphological properties of certain words in one set may be the same as those of words in another set, so that we can say that both sets of words belong to the same morphological class. Thus the N_{rel} in f^1 are similar, in respect to taking plural, etc., to the members of the operand set N, and are all included in morphological class N (as distinguished from the operand set N above). If a particular word is a member of two of these f, it has different local synonyms in each f: *He gave a book* (f^1, synonym: *handed*); *He gave a look* (f^2, synonym: *took*).

$f^1: \overline{N}_i, \overline{N}_j \rightarrow \overline{N}_i f^1 \overline{N}_j$ (*wear* in *John wears hats.*
in in *John is in bed.*
father in *John is the father of Jim.*
differ in *John differs from Jim.*

$f^1: \overline{N}_i, \overline{N}_j, \overline{N}_k \rightarrow \overline{N}_i f^1 \overline{N}_j P \overline{N}_k$ (*P*: preposition; but *and* also occurs here.
attribute in *John attributes the book to Jim.*
between in *John is between Jim and Tom.*)

possibly:

$f^1: \overline{N}_i, \overline{N}_j, \overline{N}_k, \overline{N}_l \rightarrow \overline{N}_i f^1 \overline{N}_j P \overline{N}_k P \overline{N}_l$ (*push* in *The man pushed the car from the pillar to the post.*)

In f^2:
For the set ϕ_v we have, with $i = 1, 2$:

$f^2: f^i \rightarrow f^2 f^i$ (*have-en* in *John has gone.*)
$f^2: f^i \rightarrow f^i f^2$ (*-ive to* in *John is receptive to money.*)

For the set ϕ_s we must introduce the six deformations of S listed in 4.2.2.3, and write them as numerals before the operand f. Only the main subsets are given here, with $i = 1, 2, 3$:

$f^2: f^4 \rightarrow 1f^4 f^2$ (*fact* in *That John came is a fact.*
true in *That John came is true.*)

$f^2: f^4 \rightarrow 2f^4 f^2$ (*question* in *Whether John came is a question.*
uncertain in *Whether John came is uncertain.*)

$f^2: f^i \rightarrow 3f^i f^2$ (*easy* in *For John to come is easy.*)
$f^2: f^i \rightarrow 4f^i f^2$ (*occur* in *John's returning has occurred.*
frequent in *John's returning is frequent.*)

$f^2: f^i \rightarrow 5f^i f^2$ (*slow* in *John's lettering is slow.*)
$f^2: N_h, f^4 \rightarrow N_h f^2 1f^4$ (*know* in *John knows that Jim came.*)
$f^2: f^4, N_h \rightarrow 1f^4 f^2 N_h$ (*surprise* in *That Jim came surprised John.*)
$f^2: N_h, f^4 \rightarrow N_h f^2 2f^4$ (*wonder* in *John wondered whether Jim came.*)

$f^2: N_h, f^i \rightarrow N_h f^2 3f^i$ (*require* in *John required Jim to come.*)
$f^2: N_h, f^i \rightarrow N_h f^2 4f^i$ (*instigate* in *John instigated Jim's coming.*)
$f^2: N_h, f^i \rightarrow N_h f^2 5f^i$ (*imitate* in *John imitated Jim's writing of his name.*)

$f^2: N_h, f^i \rightarrow N_h f^2 6f^i$ (*weigh* in *John weighed Jim's purchase.*)
$f^2: N_h, N_h, f^4 \rightarrow N_h f^2 N_h 1f^4$ (*tell* in *John told Jim that Tom came.*)
$f^2: N_h, N_h, f^4 \rightarrow N_h f^2 N_h 2f^4$ (*ask* in *John asked Jim whether Tom came.*)

The cases of noun dependence between parts of the argument are of the following kinds:

$f^2: N_k, f^i(N_k, \ldots) \to N_k f^2 4 f^i(N_k, \ldots)$ (*begin* in *John began his buying books.*)

$f^2: N_{hk}, f^i(N_{hk}, \ldots) \to N_{hk} f^2 4 f^i(N_{hk}, \ldots)$ (*try* in *John tried buying books.*)

$f^2: N_k, f^i(N_k, \ldots) \to N_k f^2 5 f^i(N_k, \ldots)$ (*do* in *John did his buying of books.*)

$f^2: N_k, f^i(N_k, \ldots) \to N_k f^2 6 f^i(N_k, \ldots)$ (*make* in *John made his purchase of books.*)

$f^2: N_{hk}, N_h, f^i(N_{hk}, \ldots) \to N_{hk} f^2 N_h 3 f^i(N_{hk}, \ldots)$ (*promise* in *John promised Jim that he would go.*)

$f^2: N_h, N_k, f^i(N_k, \ldots) \to N_h f^2 N_k 3 f^i(N_k, \ldots)$ (*force* in *John forced Jim to go.*)

$f^2: N_k, f^i(\ldots, N_k) \to N_k f^2 6 f^i(\ldots, N_k)$ (*suffer* in *John suffered defeat at their hands.*)

For the set ϕ_c:

$f^2: f^i, f^4 \to f^i f^2 f^4$ (*and* in *John came and Jim left*; the f^4 on the primary f^i will operate later. *because* in *John came because Jim left.*)

$f^2: f^2, f^i \to f^i f^2 4 f^i$ (*because of* in *John came because of Jim's leaving.*)

$f^2: f^i, f^i \to 3 f^i f^2 3 f^i$ (*be* in *For John to win is for Jim to lose.*)

$f^2: f^i, f^i \to 4 f^i f^2 4 f^i$ (*imply* in *John's winning implies Jim's losing.*)

$f^2: f^i, f_m^4 \to f^i f^2 f_m^4$ or not f_m^4 (*whether* in *John will come whether Jim leaves or not.*)

The first two f^2 under ϕ_c, and the last, are in the word class C (conjunctions), as can be seen from the fact that they are preceded in the resultant by an undeformed f. The other f^2 above are in the word class V (verbs), or if they are not then they receive an automatic *be* by the ϕ_m of 6.6. In all cases, the resultant of binary f^2 can contain an additional string $f^2 f^1 \ldots f^2 f^1$ (the f^2 here being binary) to satisfy the required $CS \ldots CS$ (5.6.2).

It is clear that the f^2 falls into subsets depending on their operands: primarily those whose operand is one or more N and one f, with or without dependencies among the various arguments in the operand; and those whose operand is two f (hence two S'). Each language has only a particular selection of the possible varieties of operand which the above classification permits.

In f^3 we find primarily zeroings, as in ϕ_z on ϕ_c in *He came and went* (from *He came and he went*), and permutations as in ϕ_p on ϕ_s in *It surprised him that iron is a metal* (*That iron is a metal surprised him*), and ϕ_m.

In f^4 the tense is introduced into the word sequence that has been constructed by the other f, and a sentence results. Certain f^3 operate on f^4, as in the ϕ_p: *Only now will he go.* $\leftarrow$ *He will go only now.* We can avoid producing too many f^4 in a sentence by saying that f^4 does not operate on f^3. This is made possible by the partial ordering that applies to f^3 in the analysis of a sentence: an f^3 and f^4 can both operate independently on a prior f^2. In the first two and the last f under ϕ_c above, the f^4 is concatenated not to the f^2 but to the preceding f^i in the resultant. E.g.

f^4: *John go and Jim will go* $\rightarrow$ *John went and Jim will go.*

The alternative to this cumbersome analysis is to say that conjunctions operate on f^4 pairs: and on *John went, Jim will go*. The whole operation of f^4 does not fit very well into the regularities of the f, and the situation in this regard seems to differ in different languages.

The concatenations in f^1 and f^2 are modified. It is true that each of these operators always adds something to its operand, and that what it adds is its own morphemic composition (e.g., the *begin to* member of ϕ_v adds the words *begin to* to its operand). But the added material is not always to the left of the operand; in some cases it is to the right, or between the portions of a pluriword operand. The fact that the operators consist physically of concatenations to their operands explains how it is that sentences appears physically as a sequence of words while their structure turns out to be products of operators on the ultimate single-word class N or on sounds 'q' taken as elementary entities.

As a theory of sentences, the system described above not only suffices to derive each sentence but also bears reasonable interpretation for the various entities of the theory. The arguments N and the operators f^0, f^1, f^2, f^4, all consist in almost all cases of single morphemes each; so that this analysis comes close to showing the operator contribution to the sentence of each morpheme in it. The members of f^0, f^1, f^2, f^4 all add increments of specific morphemes to their operands, and they all change the meaning of their operands accordingly, but each f makes the same meaning change in all its operands. The f^3, which act only on f^2 or f^3, are paraphrastic: they neither add morphemes (though in the case of pro-words they replace morphemes) nor do they add meaning. The analogic transformations (4.2.4) can be considered as single transformations, like the f^2; but many of them differ from f^2 in being restricted to particular subclasses of words, and all are paraphrastic. They can also be considered as products of f^2, f^3, and inverses of these; then they constitute a new and paraphrastic set of nonelementary transformations.

7.1.2.3. *Nontransformational properties*

When the purely transformational properties of the ϕ are expressed by the f operators, we see that sentences of a language have certain additional properties beyond the purely transformational ones given in the f. These properties are found to be useful for the quick processing of the f in constructing or analyzing a sentence (i.e., in composing or understanding it). Characteristically, these are the properties of language which hold almost everywhere, but not in all sentences.

One of these properties is that in most situations where two sentences are conjoined into one, one of the two (the "primary") is left unchanged. That is, $\phi_c(S_1, S_2)$ is physically $\phi_c S_2(S_1)$, leaving S_1 unchanged and concatenating to it C with a possibly modified S_2. This is one of the properties that makes string analysis a convenient method for analyzing sentence structure. An exception is, e.g., the verb-form of conjunctions as in *Their phoning caused her leaving.*

Another of these properties is that in most situations the adjoined $\phi_c S_2$ (or its residue after zeroing) is permuted to being next to the segment of S_1 to which S_2 is related, as in [*The man left*] *wh-* [*The man had waited*] → *The man who had waited left.* This makes possible the formation of constituents in sentence structure, and this too has exceptions (3.4).

A third such property is the fact that the word subsets of the various operators are like those of other operators and of the elementary sentences. For example, if we consider the ϕ_s which operate from the left (e.g., *People know that* ..., *Boys believe in* ...) we find that their first words are members of the first-word subset of elementary sentences K (*People eat bread, Boys climb trees*). As to the second words, in a few cases they are members of the second-word subset of elementary sentences (as in *I know him*). In other cases they do not appear in K; but there are certain ϕ_m which operate on the second words of ϕ_s as they do on the second words of K: e.g., the tense-affixes (in *knew, believed* and *ate, climbed*). In such ways, the set of first words of ϕ_s and the set of first words of K can be both called subsets of a superclass "Noun"; and the set of second words of ϕ_s and the set of second words of K can be both called subsets of a superclass "Verb," even though "Verb" includes almost entirely disjoint subsets of words. Furthermore, it is not only that the words of K and the words of the various ϕ are classified in the same superclasses, but also that the initial order of these superclasses in most ϕ resultants is the same as that in K, namely $NV \ldots$; and this is the portion of a sentence which is virtually never zeroed. This initial similarity of K and $\phi \ldots \phi K$ as superclass sequences makes it possible to describe briefly the effect of further operators, which make, for example, the same change in the second word of a ϕ_s (when they act on a ϕ_s) as they do in the second word of a K (when they act on a K). Indeed, unless we could formulate such superclasses we would not be able to state in a finite way how an operator acts on all of its possible (denumerably many) operands. Here too, there are exceptions: Certain ϕ_z, ϕ_p, produce word-class sequences which occur nowhere else in the grammar (e.g., the question); and certain operators introduce words which are not members of any otherwise known word-class (e.g., *yes*).

The fact that virtually all sentences are thus similar sequences of the same superclasses gives them a gross common framework which is useful for a finitary statement of how a further ϕ would operate upon the sentence, and which, like all grammatical properties, also bears a certain interpretation: one can recognize which verb is the "main verb" of the sentence (bearing tense and not preceded by a C or a portion of a ϕ_s) and one can recognize the preceding noun which is its "subject," and the interpretation is that in the sentence we are talking about the subject and asserting the main verb about it; the other parts of the sentence are then modifiers of the sentence or of its parts. Note that where this recognition in terms of immediately obvious elements fails, as in *He went home, I think* one is less sure what is the subject. The superclass string-structure of the sentence gives the assertion-standing of each part of the sentence; the transformational composition gives the meaning relation of the parts (what increment acts on what). In any case, this interpretation into subject, predicate, local and sentential adjuncts (modifiers), which is approximately that given by string-analysis, is independent of the transformational interpretation of the same sentence which follows the partially ordered ϕ and K whose traces the sentence contains. For example, the subject of *I was challenged by them* is *I;* but *they* is the subject of the K from which this is obtained by a paraphrastic transformation.

The main effect of these properties is to make a finite, and rather small, grammar serve for the computation (the constructing or the recognizing) of a denumerably infinite set of sentences. This effect is not vitiated if a reasonably small number of cases do not have these properties.

7.1.3. *A symbol-system for language-borne information*

The abstract system which suffices for language is uncomfortably complex, and this not only in a traditional analysis of language but even in a transformational analysis, and even in an analysis into base operators as in 7.1.2. Two facts, however, make it possible to improve this situation. One is that there exists a generator set for all the transformations, and that this base set is a small set of short elementary increments (and, distinguishedly, non-incremental changes) whose physical shapes can be characterized and classified in a reasonable way (as inserting V in a sentence, or adding NV to it, or adding C to two sentences, etc.). The other fact is that everything about the operators in the generator set has a reasonable informational correlation which is their constant semantic contribution to each sentence in which they appear: the superclasses N, V, etc., of which each base operator is composed, the order in which the superclasses appear in the operator, and the placing of the operator on its operand.

We can see wherein lie the possibilities for improvement if we ask to what is due the complexity in 7.1.2. It is due to several facts: that various subsets of the base operators are restricted to occurring on particular operands; that in a given set of base operators certain subsets have somewhat different physical shapes (as superclass sequences); etc. Since we can check the semantic contribution of each operator, we can see what is the effect of changing the physical shape or the domain of an operator. For example, some ϕ_s consist of NV_{-s} placed to the left of their operand (e.g., *I know that* on *He went*) while others consist of V_{s-} Ω (including *be N_{cl}*) placed to the right or around the operand (e.g., *That ... is a fact* on *He went*); nothing is changed in the interpretation if we replace the NV_{-s} by operators in the other form (e.g., placing *That ... is known to me* on *He went*, yielding *That he went is known to me* as a replacement for *I know that he went*). We can also see what is the effect of removing some restriction on the domain of an operator. Following 6.8, we can say, for example, that no interpretation is changed if we remove the restricted comparative conjunction from the grammar, replacing all *-er than* sentences by the more regular ones containing *exceeds*, and so on.

Since the decomposition of sentences into kernels and base operators gives the semantic contribution of almost every segment of each component, we can now consider how to regularize not simply the operator structure alone but also the informational interpretation of the operator structure. The methods of 5.6, 5.7 and Chapter 6, especially 6.8, can be used not merely to adjust the K and ϕ so that they are physically more uniform within each subset and have less restricted domains, but also to do this with an eye to better correlation of symbol sequences (the segments of K, ϕ) with their informational contribution to the sentences in which they appear. We try to come as close as possible to having a symbol system in which each symbol, and each way ("rule") of making sequences out of these symbols, has a constant informational contribution to the sentences in which the symbol or rule appears, and in which each sentence has a computable decomposition into these symbols and rules of combination. We would want sentences carrying different information to be different symbol sequences, and sentences consisting of the same symbol sequence to carry the same information, and sentences which differ in their symbol sequence to differ correspondingly in their information. If this latter is to apply in a reasonable way to all sentences, it is clear that the ways of combining symbols will have to fit at each point the ways of combining information.

There is no reason to think that a system representing the elements of information and their combinability can be constructed in an absolute way. Except within a well-organized science at a particular time, there is no absolute way of determining the primitive entities of information or the aspects from which they are to be regarded. However, the situation described above for natural language, when it is analyzed into the base operators, makes it possible to improve the language situation in the desired direction. In so doing, we remove dependences in the symbol sequence (i.e., restrictions on combination) which do not correlate with differences in information. We can here no longer be satisfied if the rules of combination are simple but the primitive elements are complex. Each segment must contribute appropriately to the information carried by that element; and this means that the kind of information carried by the ultimate segments must itself be only of certain kinds.

We now consider what parts of the structure of language can be eliminated or changed, for a better correlation with information.

In the finite set of N, we can ask whether there are particular subsets, determinable on absolute grounds, at least within a particular science or subject matter. In the other set of primitive arguments, 'q', we can omit the set of sounds, leaving only the set of sequences constructed out of the alphabet of the abstract system without the symbol 'q'. This leaves intact the usefulness of 'q' for forming metasentences within the language, and makes 'q' explicitly denumerable. All it loses (and only from the spoken language at that) is the ability to include nonlinguistic sounds within language.

The f^1 operators which impose inequalities on their N operands, and whose resultant is a sentence, are essential for any system that would present information; but it may be possible to restrict them to one- and two-place arguments, those with more arguments being derived from conjunctions on f^1 with fewer arguments. For a particular subject matter there may be absolute subdivisions of f^1, and subdivisions of the set N in respect to f^1.

The f^2, which preserve the inequalities imposed by f^1 while adding increments to their operand sentence, are also essential to any system which would process information. They may be restricted, as they are in language, to having only one f^1 or two f^1 in their operand, although particular subject matters may find a three-f^1 operand useful for certain f^2 (e.g., *between*). There may be many restricted f^2 in one or another language which can be extended without producing informationally undesirable sentences; this is especially likely to be the case if the operator is restricted due to lack of the necessary affixes on particular words. There are also many informationally-neutral possibilities of making the physical shapes (the word-class sequences) within a subset of f^2 more similar to each other, as in the example of the ϕ_s above. Similarly the f^0 can be changed into f^1 or f^2, by using the methods of 6.5,8. All such changes improve the correlation of symbol sequence with information.

However, just as the meaning-ranges of the f^1 depend on the inequalities which they impose on the N, so the meaning-ranges of the f^2 depend on the kind of word-classes they contribute, on the restrictions as to their domain, and on the superclass sequence which results. For example, the ϕ_v bring in only a word of the same superclasses as the f^1 (i.e., predicational words such as verbs), and place it in the position of the operand f, as in ϕ_v: *They go → They begin going, They are ready to go*. Hence ϕ_v can only modify the predicational part of their operand.

In the ϕ_s, there are some which operate only on f, and are placed in respect to that f just as that f had been placed in respect to its operand in in turn, as in

ϕ_s: *He returns → His returning is important, His returning is a fact.*

Such ϕ_s are necessarily predicates on their operand f. There are other ϕ_s which operate on a pair N, f, as in

ϕ_s: *He returned → I know that he returned.*

These necessarily relate a noun to a sentence (an event, etc.) just as a two-place f^1 relates two nouns to each other. If the noun is restricted to the human subset, the ϕ_s can only indicate the relations that a human can have to an event. And so on. New situations also arise, which can only carry certain new kinds of information, For example, there are some ϕ_s which operate only on disjunctions of f, as in

ϕ_s: *He returned → I wonder whether he returned or S ... or S.*

Such ϕ_s can only express the relations that the human can have to a disjunction of predications, which is an uncertainty in respect to any one of the predications. When the disjunction f_a or f_b or ... or f_n is paraphrased in respect to a particular f_a, as f_a or not f_a, the uncertainty is specifically predicated about the f_a : *I wonder whether he returned or he did not return.* If the ϕ_s is restricted to a particular domain of operands, its meaning is restricted correspondingly. Thus the adverbs of manner, which operate only on those f which are in the verb superclass (i.e., not on the pseudo-verb *be* of *He is sick, He is a man*) carry a kind of meaning which can be predicated only of real verbs.

The semantic difference between the ϕ_c members of f^2 and the others arises from (or is expressed by) the fact that ϕ_c operate on pairs of f^1. The difference within ϕ_c between the *and, or* of both language and logic and the *because*, etc., which are available only in language arises primarily from the fact that in the latter the resultant depends on the words of the operand :

$$because\ (S_1, S_2) = S_1\ because\ S_2\ and\ S(1, 2)$$

where the $S(1, 2)$ required in the resultant is a sentence which connects the words of the two operand sentences under *because*. The semantic effect of this is to make the sentence-connective depend on the connection among the words of its operand sentences, and it is this that makes a substantive as against a set-theoretic connective.

In this way the kinds of symbol sequences which each f^2 produces out of its operand determines the kinds of meaning the f^2 can carry. This can be used in deciding what changes we can make in the f^2 of natural language without changing the meanings available, and what changes we can make in order to obtain specified changes in the meanings which the symbol sequences can express.

However, there seems to be no informationally-useful way of eliminating the whole inner set of ϕ-ψ (4.3.1, the ϕ which do not participate in the semi-group of ψ). The meanings which these base operators carry are in many cases such as can operate only on some but not all K or ϕ. Thus, even if a grammatical generalization can be formulated that would extend the ϕ_s of manner to the *be*-sentences (such as *He is a man, His breathing is heavy*), there might be no useful meaning to attach to the resultant (e.g., *His being a man is slow, He is a man slowly*). In a specified subject matter it may be possible to specify certain subsets of ϕ which are restricted to operating only on specified subsets of K, ϕ. There may even be subsets of ϕ which impose inequalities on their operands somewhat as the f^1 do on the N. The main example of this in language as a whole is the ϕ_s of manner (*He breathes heavily* is more acceptable than *He looks heavily* or than *He breathes darkly*). Such operators should perhaps be looked on as not f^2 but an outer set $f^{1\prime}$ of f^1 operating optionally on the regular f^1. This would be equivalent to defining a macrokernel $NV\Omega D$ in which the optional D represented the adverbs which fit the particular V.

The transformations which make no contribution of information, but leave the meaning of their operand unchanged, would of course be eliminated. What parts or changes in the symbol sequence are to be thus eliminated is readily specifiable, because transformational theory can distinguish the paraphrastic transformations from the others: they are the base operators f^3 which do not operate on K (permutation ϕ_p, zeroing ϕ_z, morphophonemics ϕ_m), the analogic (irregular) ϕ (4.2.4, such as the passive and the middle), and the synonymy replacers (6.5, including all reducers of vocabulary such as the replacement of *wh* and other conjunctions by *and* with suitable CS_3). The synonymy replacers are stateable to only a limited extent in language as a whole, but may be more definite for particular subject matters. It must be remembered in this connection that the kernel and incremental-ϕ composition of a sentence is left unchanged by the paraphrastic ϕ, so that elimination of the latter requires no special checks.

The f^4 can be replaced by the domain-conditions of the f^1 and f^2. These conditions specify which sequences of f^1 and f^2, and of the primitive arguments of the f^1, occur in the system. The f^4 indicate the main verb of each such sequence, or of certain segments of it; but this can be stated as an interpretation of the sequences. The tense information carried by the f^4 and by tense-adjuncts on the f^4 would be replaced, in the manner of 6.8, by free ϕ_s of time on each f or on only certain f.

It is thus possible to replace the system of 7.1.2 by a simpler system, possibly an ordered triple $\langle N, f^1, f^2 \rangle$: the N being a finite set of primitive arguments; the f^1 a finite set of operators on N or N-pairs or sequences written in this alphabet (these are the 'q'), each f^1 imposing inequalities on its operands; the f^2 a finite set of inequality-preserving one- and two-argument operators on $f_1^2 \ldots f_n^2 f^1$ (for $n \geq 0$), in which one subset contains families of operators each with a restricted domain of operand, while in the complement subset $f^{2\psi}$ the products form a semi-group. The interpretation is that each f is an assertion (a predicate) about its operand, even in the case of the two-place f^2 (i.e., the conjunctions).

Many systems, of various complexities and interpretations, can be made in this way, depending on how the families of members in each set interrelate in respect to their physical composition and their domain. If we want the system to support approximately the interpretation of natural language, we have to start with the subsets that exist in natural language, and modify them either so as to improve the correlation between differences in physical shape or domain and differences in information, or so as to alter the interpretation in predictable respects.

Some difficulties which exist in language can be eliminated from the new system. An important case is that of ambiguous sentences (i.e., word sequences which represent more than one proposition), which are found in every language. Many occurrences of ambiguity are avoided by eliminating ϕ_z (zeroing, which is the main source of ambiguity) and the other paraphrastic ϕ (especially the analogic ϕ, which require a similarity between two differently-constructed word-class sequences). Other ambiguities are eliminated if we require that no one constant be a part of two or more different base operators (e.g., that the morpheme -*ing* be part of only one base operator). Finally, while we may not want to lose the advantages of having superclasses (above), we can prevent them from leading to ambiguities by requiring that their subsets be disjoint. This condition is needed only for the characteristic words of f, in effect their verbs and adjectives and prepositions and predicate nouns. Then each family of f^1 or f^2 would consist of words unique to that family: e.g., the verbs of the kernel would be different from the verbs of ϕ_s. This is largely the case in natural language, but there are many exceptions (often of metaphoric origin) and it is these that supply the ambiguities.

No system, however, is practical without some method of abbreviating its sentences. This includes a method for reference, i.e., for making it possible to indicate when various word-occurrences in a discourse refer to the same individual, without having to identify each occurrence by counting its position in the discourse. Furthermore, since all zeroing in language applies to material which is reconstructible from the environment, and hence does not carry information in that environment, an efficient symbol system for information should provide kindred abbreviations. The only undesirable feature of all these abbreviations is degeneracy, when two abbreviations, in two different sources, produce the same word sequence in the same environment. Hence it would be necessary to formulate abbreviation and reference rules more restricted than those of language, in order to avoid degeneracies.

We can now ask what kind of simplified system we would be getting from the triple described above. It is not an absolute system for information, though it may be partially so for particular subject matters. It is not natural language, since it lacks various conveniences and inconveniences of language, and lacks the internal provisions for change. Rather, it is a better-correlating symbol-system for the information which is borne in language (since we can check that each improvement in the symbol sequences did not change the information carried), or for a type of information which is modified in explicit ways from that borne in language.

The system would consist of entities, called sentences or expressions, each of which is a partially ordered set (which may be written as a sequence, parts of it commutative) of N, f^1, and f^2 satisfying the conditions stated above. And since information is given chiefly in whole discourses, which have the properties of 5.8, the system would consist of discourses, each of which is indeed a sequence of sentences but with each sentence immediately followed by certain modifications of it.

7.2. Homomorphisms and subsets

The set of sentences takes part in certain homomorphisms preserving the transformational relation (Chapter 5). The set consisting of both sentences and infrakernel sentences maps homomorphically onto the set of sentences. The set of sentences and unambiguous sentence pairs maps homomorphically onto the set of sentences. The set of sentences maps homomorphically onto the set of paraphraseless sentences (omitting all sentences produced by f^3 or by analogic products involving these); in this set there are no ambiguities or grammatical paraphrases, but the *and, or* binaries become commutative operations: $aRb \equiv bRa$. (In effect, aRb and bRa are here paraphrases.) The paraphraseless set (and also the whole set of sentences) maps homomorphically onto the synonymless set of sentences (where all synonyms of a basic vocabulary of words, or of sequences of these words, have been eliminated by certain ϕ_m which entail adjoined sentences that state the synonymity). In this last set of sentences it seems (although it has not been fully shown) that each member of f^1 gives value 1 of acceptance (normal) for a unique set of N values, and perhaps each member of f^2 gives acceptance 1 for a unique set of f on which it operates. Loosely speaking, each word in the language enters into a unique range of combinations with other words, within f-defined segments, in sentences of normal acceptability. These differences hold also in the whole set of sentences, but are clouded there by the difficulty of recognizing the f segment in which the given word was placed.

There are interesting relations among the mappings which have been mentioned in preceding chapters. Thus, we take S_ψ as the set of ψ-decomposable propositions, on which are defined the lattice-like ψ-structures representing the ψ-traces in each proposition. E_ψ is the set of elementary propositions in S_ψ. Then we have a short exact sequence of the mappings

$$O \rightarrow E_\psi \rightarrow S_\psi \rightarrow S_\psi / E_\psi \rightarrow O,$$

where S_ψ / E_ψ is the monoid of lattice-like ψ-structures, and $S_\psi \rightarrow S_\psi / E_\psi$ is the natural mapping mentioned toward the end of 4.3.2.

The set of sentences also has certain characteristic subsets, which are identifiable in respect to transformations. Particularly important are: The two-place f^1 which state that a word or sequence x is a synonym of a word or sequence y under stated f; the metatype f^1 which state that a cited segment in certain sentences, possibly in a stated neighborhood, is of a particular class; the metatoken sentences which state that the pair of a position and a segment (i.e., an occurrence of the segment) in a cited sentence or discourse has a particular property; and the subject-matter sublanguages (e.g., of science).

7.3. *Essential properties of the description*

It seems that all languages are similar in having the structure described here, and it is therefore of interest to ask what properties are essential to language in terms of this structure. Only two bodies of data are used:

1. Phonemic distinctions in sound sequences (cuts in the set of sound occurrences), i.e., the results of the pair test (3.1), even though modified by later regularization. The set of all phoneme sequences includes, as a proper part, the set of utterances in a language (with allowance for utterances that contain nonlinguistic sounds).

2. Inequalities of acceptability or normal discourse-neighborhood for phoneme sequences as utterances of the language. This body of data can be used in two parts: the set of all phoneme sequences which have nonzero acceptability may be used in determining morpheme boundaries; and the inequalities within this set may be used in determining transformations.

While any particular bodies of data of these types determine a particular language, different languages have different data but of the same type. For a given language, if all the data were replaced by other data whose satisfaction of the two conditions above were one-to-one to that of the existing data, the result would clearly be a language equivalent to the given one except for the physical difference in the data. Indeed, in the case of writing, the phomemically distinct sounds are replaced (only approximately, at that) by letters, and the result is an equivalent language of marks instead of sounds. The physical objects are thus seen to be arbitrary in respect to the structure. The structure of a language is characterized by how an arbitrary set of objects satisfies the two conditions above, and by how the resultantly defined objects fit into the abstract system presented in 7.1.2.

Since the determinability of the elements of a science is no less relevant than the determinability of the operations or rules on these elements, we now consider what is the complete description, from data to sentences.

The most general property is that no description of language goes in one operation from data to the determination of which sound sequences are sentences (or from phonemes to the determination of which phoneme sequences are sentences). In string theory we have the following steps:

1. from sound sequences to phoneme sequences by the pair test;
2. to morpheme boundaries by the recurrent hereditary process of 3.2;
3. collecting morphemes into classes, forming elementary strings of these classes, collecting these strings into sets having the same entry properties, all in order to satisfy:
4. a recurrent hereditary process on morphemes (or words) in terms of their string position, yielding the sentence boundary of those morpheme sequences which are grammatically possible sentences;
5. an inequality test as to the acceptability, or normalcy, of the particular grammatically possible sentences.

In transformation theory, we have the same first two steps, then in the system of 7.1.2:

3. the collecting of morphemes into the domains of the various symbols of the system: into the primitive N, into each subset of f^0, of f^1, of f^2, and of f^4;
4. the defining of the conditions for operation of f^0, f^1, f^2, f^3, f^4 (the ordering being introduced under f^1); the resultants of f^4 being sentences, on all sequences containing f^1.

In the form used in Chapter 4, of transformations in the set of sentences:

3'. the collecting of morphemes into classes, such that in certain n-class sequences the n-tuples of word choices are organized into inequalities as in the second set of data above; and the collecting of other morphemes into the subsets of transformational traces;
4'. the defining of the n-class sequences of 3' as elementary sentences, and of the base transformations and products of them as a way of deriving sentences from sentences.

It is clear, then, that no method is available for characterizing sentences directly in terms of some property of its indecomposable elements (whether transitional probability of its sounds, or some structural classification of its phonemes, or whatever). Each of the four or five steps in each list above can be described as an independent structure consisting of elements, operator, and resultants, with the resultants of each nth-level structure being mapped onto the elements of the $n + 1^{th}$. The fact that everything in a sentence is sounds, or words, in a sequence is merely a physical description of the sentence. In terms of linguistic structure, the words in a sentence are the physical form of operands and operators, the argument (within the sentence) of each operator not being marked off by parentheses only because we can recognize from each word what operator subset it is classified in and how many words of what subsets are its argument.

An overriding property of all languages is that while they seem to contain simply sequences of one class of objects—sounds or words—these sequences turn out to be members of operands and of operators acting on them.

This result is achieved by the fact that the operators themselves consist of words, or changes in words, and that the words of the operators are merely concatenated to their operands: i.e., the operators consist of words and are contiguous to their operands. A further concealment of the operator structure is due to the fact that many of the operators (but not affixes or conjunctions) consist of words which are of the same superclass as their operands, i.e., have certain similarities in structure and further operability to the words of their operand. The reasons for all this follow partly from the properties of Chapter 2, and partly from considerations discussed below.

There are two kinds of situation in which a property common to many sentences has what appear to be exceptions, i.e., where the property seems to hold almost everywhere. The first has to do with the fact that in addition to the essential properties of sentences, there are other properties which may be essential for the language as a whole but need not apply to each sentence. These are involved primarily with the condition that a language has to have a finite and reasonably small metatheory: i.e., the instructions or habits sufficient for human beings to learn and use a language must be of manageable number and size. This may impose certain restrictions on the denumerably infinite set of sentences, which must be recursively describable by a finite grammar; but a finite subset of sentences could disregard the restriction without making the grammar infinite. Examples of the major exceptions of this type:

1. The statement that for every sentence there is some sentence to which it is transformationally related may have a finite and small number of exceptions: e.g., *Hello!*; possibly some grammatically petrified proverbs.

2. The requirement that the resultant of a transformation be similar to an elementary sentence is necessary, to avoid a denumerably infinite set of arguments for transformations. The transformations are defined on the elementary sentences, which have for the most part the form $NV\Omega$; furthermore, the transformations mostly involve changes or additions in

the NV .... If also the resultants of transformations have the form NV ... (even though the subsets of N, V may be different from those in the elementary sentences), then the transformations defined on $NV\Omega$ can easily be extended to apply to the resultant of any product of transformations.

This situation explains also why languages have their word classes collected as subclasses of major classes; for if there were not some common properties which hold both for the first word of elementary sentences and the first word of many transformational resultants—the properties of taking plural, etc., which apply to the superclass N—we would not be able to say that elementary sentences and transformational resultants are similar in having some subset of N as their first element. In any case, this requirement of similarity is suspended for a finite and small number of transformational resultants, e.g., in the ϕ_p which produce *This I like*; the form, NNV, is not like that of any elementary sentence.

In addition to the similarity of resultants to elementary sentences, the analogizing transformations have operated in such a way that most increments can be transformed from their original form to that of the others. Thus many ϕ_s appear also in the form ϕ_a (e.g., adverbs of manner, 4.2.2.1), many increments are both in ϕ_v form (by ϕ_z on identical subjects) and in ϕ_s form, many C_s appear as verbs between deformed sentences (*He came because she left.* → *His coming was because of her leaving.*) All this makes the word-class sequence of sentences more obvious than the operator subsets, though only the latter affords an adequate analysis.

3. There are other properties which hold almost everywhere and which apply not to the finiteness of the rules but to the ease of their computability. Such is the permuting of $C_w S_2$ to immediately after the N_i (in S_1) which is common also to S_2 (*The day arrived that we had long awaited* → *The day that we had long awaited arrived*); this is the frequent but not universal propinquity of adjuncts which made constituent analysis serve for much of the language but not for all. Another example is the avoidance of intercalated conjoinings of sentences, which nevertheless occurs in such isolated forms as with *respectively* (*He and she play violin and piano respectively*).

The second type of exception has to do with the fact that language changes and that there are always possibilities for new sentence forms. The possible innovations are, however, not arbitrary objects—if they were, they would not be recognized or understood as sentences of the language—but extensions of the domain over which particular transformations are defined (or minor variations of the physical composition of the transformation). If it becomes possible to predict the kind of extensions and variations, on the basis of the current definition of the transformation and of its relations to other transformations, we could go further than the present statement that the exceptions are only extensions: we might be able to delimit the possible extensions of the language.

The systems discussed here can be so devised as not only to achieve a compact system from which the facts of the language can be derived, but also to eliminate the all-too-frequent recourse to a mass of individual facts (idioms, exceptions, etc.) which lie more or less outside the grammar. The subsets of operators can be so defined that many peculiar forms and relations come out as resultants of particular products of elementary transformations (e.g., the transformability of *He spoke hastily* and *His speaking was hasty*; the lack of adverbs of manner on *is, costs*, etc.; the unacceptability of *He frequently and slowly wrote these letters*).

The systems can also be so devised as to make it unnecessary to add outside the theory any general statements about properties of language. This is achieved by using every known property as a basis for reducing the system itself. For example, the linearity and discreteness of language material were used for counting it, in the interest of reference, etc. (5.6, 7); the inclusion of the metalanguage was used for reducing conjunctions, synonyms, etc. (5.6, 7; 6.5); the fact that different words or markers (morphemes) appear in different operator subsets makes it possible to identify the occurrences of the operator in the sentence as a correlation between the syntactic definition of the operator and the morphemic membership of the operator.

7.4. Languagelike systems

We can see the importance to natural language of the various operators and of their relations by considering the effect of removing each of them. This will also give some indication of a classification of systems which are partially similar to language.

If a system lacks f^1, its elementary sentences (no longer defined as $f^1 N$ but as sequences of undefined objects—words) become primitive objects with respect to the system, with identities and similarities determined only by physical (e.g., phonetic) composition, or by coordinates in a space (e.g., position in a sentence).

If a system lacks unary transformations it can only deal with objects (f^1 on N) but not with events, facts, statements: it would lack the grammar and vocabulary for modulations (e.g., beginning), time properties (e.g., frequency), manner, etc., of action; for initiating events; for knowing facts, etc. This is so because the meanings of words of f^2 cannot be carried by words of f^1. Languages (or sets of sentences) dealing with restricted objects of the world could dispense with the unaries, or at least with the ϕ_s. Discourses dealing with events and the like would need certain subsets of ϕ_s but not others. Languages containing their own metalanguage must have certain f^1 (is N_n), and related ϕ_s, which operate on sequences in the alphabet.

If a system lacks the binary transformations, ϕ_c of f^2, it cannot construct relations among sentences. If it has binaries but without the word-repetition property, it would not be able to express substantive relations between events; the word-repetition condition may be syntactically equivalent to the introduction into each sentence of the acceptance ordering of the relevant words, i.e., their dictionary difference (Chapter 8). Material implication, in logic, does not require the word-repetition property; causation does (and this in turn requires that the language contain in normal acceptance the zeroable sentences necessary to fill out the word repetition).

If products of the f^2 were commutative, all the f^2 would apply without order to the f^1 in each $f^2 \ldots f^2 f^1$. One could not operate on an f^2 as one operates on an f^1: one could begin an action or know about an action, but one could not begin to know or know about beginning. As to associativity, the binaries whose iteration is associative and which are (in the paraphraseless set) commutative, e.g., and, or, are available in logic as well as in natural language, and are semantically much weaker than the other binaries. If inverses of ϕ occurred freely as transformations, the set of transformational products would have symmetries that would conflict with the amount of distinguishing of information that is needed in language.

If the transformations of a system were not partial but were defined over the whole set, there would not be possibilities of extension of the domain of an operator. There would also be no possibilities for analogic transformations, which require in addition to this also the use of inverse transformations (generally following a zeroing), and which entail that not all products of base operations be transformations. A system without all these conditions would be not open, like natural language; it would be unavailable for any internal extension, and would be suitable only for a closed language codifying certain fixed information.

A system in which all members of f^1 had normal acceptance for identical domains, i.e., in which there were no acceptance inequalities, could not express meaning differences among the members of the operator set. This is the case in logic and mathematics. The redundancy that remain in the synonymless and paraphraseless set of sentences is essential if the words are to have different meanings (see Chapter 8, fn. 3).

As to the physical properties: In a system in which the resultants of operators preserve the general structure of some already defined operands, i.e., in which the well-formedness of operands appears also in the resultants, an operator defined on one set of operands can be readily extended to operate on all resultants which satisfy the same well-formedness. Thus a finite grammar suffices to describe recursively the unbounded products of f^2. Without the preservation of well-formedness, the effect of each operator would have to be defined separately for each resultant on which it acts. The grammar could then not be finite.

In the matter of contiguity: A system richer than language, one whose objects (elements and their sequences) occurred in a defined space or format, could dispense with contiguity of operator. This is the case in music, where a segment of a given sequence (e.g., its last note) can be put at a given distance (a given position of a given bar) away from the rest of the sequence. In natural language, this is impossible because the distance could not be measured or named (until other operators had filled it in).

A system poorer than language would be one on which the contiguity was defined in terms of words, i.e., in which the operators had to be contiguous to particular words in their operand. In natural language one is not restricted to this, for one can find certain strings of words (derived from the elementary sentences) such that any operator related to any of the words in that string leaves an effect which is contiguous to the string, although not necessarily to the distinguished word.

E.g., the plural of a noun leaves a mark not only on the noun but also on the verb of which that noun is subject: the plural operates on the noun–verb string, not only on the noun. In the case of adjuncts (modifiers) A of the noun N_1, we say not that A is an operator on N_1 but that a string containing A and N_1 (and stating the position of N_1 in Z) is an operator on the string Z containing N_1, and that an idempotent operation (a case of zeroing) eliminates the N_1 from the operator. This permits adjuncts of N_1 to occur at a distance from N_1, while remaining contiguous to the string of which N_1 was part, as in adjectives in Latin, or in *My friend came, whom I had told you about*. Defining the operands as strings overcomes some of the limitations due to having no space in which the sentence is located: it enables a word to have certain relations to certain other words which are contiguous not to it but to the string of which that word is part. A system in which such strings could not be defined, and in which the operator could be defined only as operating on the individual element (here, the word), would not be able to have modifiers, grammatical dependencies (" agreements "), and the like at a distance from the operand word. This would limit the possibility of inserting further material into a sentence.

A property common to language and to formal systems, the fact that the elements of each language event (each speaking or writing) are ordered, is essential to reference, more precisely sameness of reference (cross-reference), i.e., to noting that a given word occurrence refers to the same object as another word occurrence. No system in which occurrences of the elements are unordered could have cross-reference.

The possibility of stating metalinguistic sentences within the grammar of the language makes the language describable within itself. And the fact that these contain metatoken sentences, as well as metatype, makes reference describable within the sentence itself. Otherwise, the description and interpretation of the sentence structure and of references in the sentence (pronouns, etc.) would have to be done in a separate metalanguage which would in turn have to be defined in a separate metalanguage of it, and so on without end.

In all of the properties above, any alteration produces a system different from natural language. In contrast, the homomorphisms of the set of sentences mentioned in 7.2 produce systems which are equivalent to natural language in information-bearing. The set of unambiguous subsets carries the same information as the set of sentences. So does the paraphraseless set, except that here certain abbreviation of sentences (in particular of $SCS \ldots CS$) becomes impossible. So does the synonymless set, which has the property of minimum redundancy for the information which the language can carry. And so does the set with added infrakernel sentences, which has the property of greater regularity of operands for the base operations.

7.5. *Language compared with logic and mathematics*

The specific description of the properties of language makes it possible to see the main similarities and differences between natural language and logical or mathematical systems. It is clear that the difference lies not in any impossibility of a precise description of natural language: the synonymless set under the incremental base operations can be described precisely, even though the detail necessary for a complete description would still be prohibitive. Nor does the major difference lie in the existence of linguistic ambiguities. It is true that every natural language has ambiguous sentences, so far as we know. But it follows from 7.1.3 that a system expressing the information borne in language can be constructed without ambiguity. Hence ambiguity is not necessary for the kind of information carried in language as against that carried in logic. The difficulties of comparing the two types of system are lessened since it can be shown that questions, imperatives, and other such sentences which are beyond the scope of logic are transforms of assertions (*Come!* ← *I request of you that you come.*), so that the paraphraseless set of sentences contains only assertions.

Both natural language and mathematical systems have variables which can take values in some domain, and well-formed sequences of these variables (and constants) as elementary assertion-forms. The great difference is that whereas in logic and mathematics a certain set of elementary well-formed sequences is mapped, for all values of its variables, onto the two values T and F (or onto a more complex system of values), in language each of the elementary well-formed sequences of n variables is partially ordered by the n-tuples of values of its variables; or we can say that the set of n-tuples of values is mapped onto the set a of acceptance values, $0 < a \leq 1$. It is this partial ordering which gives meaning to the sequences (elementary sentences) and to the values of the variables (i.e., the individual words in a class) in language, whereas logic and mathematics deal only with truth value. The preservation of this partial ordering by linguistic transformations gives assurance that these preserve the meaning of the elementary sentences, aside from any meaning that their own morphemic increment may add.

The next major difference related to this is the existence in language of a number of subordinating connectives C_s which require that the S_1, S_2 which they connect be part of a longer (but in part zeroable) sequence $SCS \ldots CS$ in which each value of the variables in S_1, S_2 occurs in at least two S. This assures that there shall be a substantive connection between S_1 and S_2, since the required word repetition provides a chain of meanings connecting the words of S_1 with those of S_2. Although the analysis of these binaries in language has not been completed, it may be hoped that we will be able to say that if S_1 and S_2 have certain similarities and differences with each other, then, e.g., S_1 *because* S_2 is not nonsensical (although it may be false).

This contrasts with material implication, the connective in formal systems, which by its definition can only suffice to preserve the truth of S_1 in S_2, but cannot serve for any substantive connection of meaning between them.

It follows from this that in language the effect of certain modalities on implication is obtained not by an operator on $S_1 \supset S_2$ but by conditions that have to be met by the word difference between S_1 and S_2.

The paraphrastic transformations are somewhat like the abbreviated or expanded notations in logic and mathematics, as contrasted with a normal form for each expression. However, the unary incremental transformations, ϕ_v and ϕ_s, introduce modifications and operations upon the content (meaning) of the operand sentence, of a kind which is unavailable in the formalism of logic and mathematics (although a small part of this is parallelled in probability theory). Language also has, differently from logic and mathematics, a sentential form -*is* N_n (a subset of f^1) which creates metalinguistic sentences inside the language. Taken together with the provisions for *CS*, this means that a sentence S_i can carry CS_{meta} which make metalinguistic classifications of it and its parts. This makes it possible to use the linearity of language for identifying, in the CS_{meta} attached to S_i, the parts of S_i, and so to use this CS_{meta} for making cross-references and for assuring identity of reference; such effects are obtainable in logic only by placing the affected propositions within the scope of the same quantifier.

Between the two systems there are many contrasts which can be studied in terms of the analysis presented here. For example, the cross-referencing work that is done in logic and mathematics by variables is done in language by pronouning and zeroing. And where in mathematics derivations permit one to judge the truth of a statement from that of its premises (and classically the falsity of the latter from that of the former), in language they permit one to judge the meaning of a statement from that of its premises.

7.6. *Types of mathematical structure in language*

We can summarize what has been found so far about mathematical structures in language. It should be clear, however, that this does not refer to restricted subsets of sentences or grammatical relations but to properties of the whole set of sentences. And it does not refer to sets which only intersect the set of sentences or include it as a proper subset. For example, it does not refer to the set of word sequences: there are sentences which are not word sequences (those which contain nonlinguistic sounds of the class 'q'); and there are word sequences which are not sentences; and there are word sequences which map onto several grammatically different sentences (namely, the ambiguous word sequences).

Recurrent hereditary process. The set of all grammatically possible sentences was obtained as a subset of all phoneme or letter sequences, by two successively applied processes, the second applied on the outcome of the first. In each of these we define a possibility space for the finite set of possible outcomes (i.e., the next phoneme; or the next morpheme) but disregard the probability weights for each outcome; i.e., we deal only with the outcomes that have any positive probability. We take the possibility space at the nth point as a function of the outcomes at points

$1, \ldots, n - 1$. Certain phoneme (and, later, morpheme) sequences are then found to contain recurrent points after which the earlier events have no effect on the outcome. These are utterances in language, and the recurrent points are the boundaries of morphemes and of sentences, respectively.

Redundancies. Each restriction in the successive statements needed in order to characterize language, beginning with the pair test and ending with discourse analysis, introduces a redundancy into language. It is characteristic for language that not only is there a large total redundancy, but that this is built-up out of a number of restrictions, one operating on the other.

Transformations. It was seen that we could define a set of transformations, some of them partial, from the set of sentences into itself. The transformations are based on an equivalence relation in the set of graded sentences, and induce a partition on the set. Every sentence, for each unambiguous grammatical meaning of it, has a unique decomposition, via these transformations, into elementary sentences. The transformations can also be looked upon as operators on the set of sentences into itself, or as a special set of prime sentences such that each sentence has a unique factorization into these and the other elementary sentences mentioned above. The transformations themselves are products of a set of base transformations. The set of sentences has various subsets in respect to the base transformations; one type of these subsets (the subject-matter subsets) is a sublanguage.

Markov chain. A major advantage of the unique base-transformational decomposition is that language is characterizable by a Markov chain of ϕ (or f), whereas no Markov characterization of language in terms of words or nontransformational entities is possible.

Enumeration. Every sentence is finite, and one might first think to enumerate the sentences as sequences of phonemes or words. However, some word sequences map onto several grammatically distinct sentences. In any case an enumeration of word sequences is irrelevant. In contrast, the structures discussed above make possible an enumeration of the contribution of each structure to the set of sentences, which reveals various properties of various interesting subsets of sentences. It is found that there are various finite subsets of importance—the elementary sentences, the maximally classificatory metalinguistic sentences (i.e., the finite grammar), the elementary lists of exceptions (including nontransformable sentences, such as *Hello.*). The remaining sentences are obtained recursively by iterations of certain of the classes of transformations. In order to have all denumerable material in language come only from unbounded iterations of operators, it is necessary to obtain the words for

the numbers from a finite word set (containing only *one*) and from iterations of *and* on sentences containing the word *one*.

Recursivity. It follows from the above that the system of K and ϕ (or of
N and f) is finite and characterizes a recursive set of sentences. However,
additional rules may be needed to state that particular other sequences of
words of phonemes are also sentences. The additional rules, and sentences,
would certainly be enumerable, but it has not been shown that they must
be recursively generated.

Types of operation. Unary and binary operators had been found on the
set of sentences. Some of the binary operators are commutative and
associative (the ones which are also found in logic: *and, or*) in respect to
their paraphrase-sets; the others are neither. Inverse operations exist only
in a restricted way.

Ordering. Inequalities and ordering appears at several points. The
primitive set N is ordered by each f^1.[5] We can obtain an uninteresting
partial ordering for all sentences by considering the arguments of each f_i^1
to be noncomparable with the arguments of every f_j^1, $j \neq i$. The inequalities of the ordering under f^1 are used in defining f^2, f^3; but the transitive
property is not needed in this definition. The decomposition of each
sentence into transformations and kernel sentences (or into prime sentences) is partially ordered, and in particular can be arranged to form a
nonmodular lattice. As to linear order, it appears above all in the sequence
of phonemes or letters, and in the morpheme segment, word and sentence
boundaries which result from the processes of 3.2, 3.6. String entry points
(3.5) are linearly ordered in a sentence, and so are the locations of transformational traces (which can be looked upon as first the concatenation
of the trace with its operand, 7.1.2, and in some cases the permuting of the
trace to some other point of the operand).

Types of sets. As to the major types of sets closed under the linguistic
operators we have: the set of sentences constitutes a groupoid under the
(nonassociative) binary operators; the set of transformations which are
products of the ψ transformations (without ϕ-ψ or the analogic ones), is a

---

[5] Starting from this we obtain an ordering for the f^1. We form for each one-place
f_i^1 the set $N_{acc(i)}$ consisting of those N for which $f_i^1 N$ is maximal (i.e., the list of normal,
highest acceptability, N for that f_i^1). It can be shown that for each pair f_i^1, f_j^1 there
is some f_k^1 such that $N_{acc(k)}$ contains $N_{acc(i)}$ and $N_{acc(j)}$ and is the smallest of the N_{acc}
sets which does so. There is an f_p^1 whose $N_{acc(p)}$ serves for the universal element; this
f_p^1 is the disjunction of pro-predicates: *do it, is such.* There is an $N_{acc(0)}$ which is contained in every N_{acc}, and so serves for the null element: this $N_{acc(0)}$ is simply the disjunction of pronouns (*he, she, it*). The normal subjects of the predicates thus form a
lattice, the pro-predicates having all subjects, and the pronouns being subjects of
every predicate.

semigroup or a monoid, if we define an identity transformation; the set of words formed in the alphabet of string-relation symbols (3.7) is a quasi-group (with right and left inverses which are not identical) such that the set of grammatically possible sentences is the set of all words which cancel to zero (i.e., the kernel of the mapping of the set of all words onto the set of reduced words). There is no inverse operation in respect to which any of these structures is closed; no group of elements generates precisely the set of sentences. This is because the set of sentences contains no large symmetries (except for those which arise from the partition induced by the transformations), something which is not surprising in view of the fact that symmetries do not contribute to information.[6]

Mappings. There are a great number of interesting mappings among the defined sets. There are mappings of the resultants of one operator onto the elements of the next (7.3). There are mappings between the set of sentences or subsets of it and other subsets. These serve as the basis for defining the essential properties of sentences, such as the transformations, or various special applications, such as the unambiguous versions of each sentence. The relations among the mappings have hardly been investigated as yet. There is clearly much to study about the mappings, both in order to define specific sets that contain or are contained in the set of sentences, or have a new structure, and in order to see how the relations among the mappings can lead in general to new extensions or embeddings of the sets defined so far.

However, all of the structures as they now stand are of very limited mathematical interest. They are insufficiently regular, and in some cases have disturbing constraints. The mathematical interest may lie in specifying what are the essential points that make these structures depart from their nearest neighbors within mathematics, and how these essential disturbances are related to the semantic burden that natural language alone can carry.

---

[6] On the whole, the sets which are of interest in mathematical linguistics have only one operation. It is possible to try to construct a set with two operations, using ϕ_c as one and defining a second out of the unary transformations.

8

The interpretation

The relevance of the rather cumbersome abstract system given in Chapter 7 is not merely that it can be defined, and that it covers the whole language, aside from the limited continuous phonetic phenomena and the literary subtleties of style, allusion, etc. What is important is that it correlates with many semantic and applicational features of language, and indeed that natural language can be understood as an interpretation of this abstract system.

In the first place, each structure has its own subclasses of words, especially in the synonymless form; so that it is not as though each structure was simply a different utilization of the same words, which is being offered as a model with empirical procedure. Quite the contrary, a first approximation to the various operators can be obtained as a correlation of different words with different positions in reasonably short sentences; this can then be corrected on purely syntactic grounds (operator-operand distinction), to allow for identical words as values of two different structural symbols (e.g., *know* as V of elementary sentence and also as verb of ϕ_s). Thus the argument N of the kernel sentences are in general simple concrete nouns, in most cases single morphemes: e.g., *house, book*. The verbs and adjectives of the one-place f^1 (e.g., *exist, large*) are for the most part different from each other and from the verbs and adjectives of the two-place f^1 (e.g., *see, same*). The nouns of the one-place f^1 are classifiers and the like: *mammal*; and the nouns of the two-place f^1 are relational: *synonym, father*. The P of the one-place f^1 are a few locational prepositions: *down, here*; and of the two-place f^1 almost all prepositions. The V, A, N of f^2 fall into many different subclasses, each with particular properties (modifications of the argument sentence, etc.), and each with different semantic character: ϕ_v: *have-en*; ϕ_s: *know that, wonder whether*; ϕ_c: *because*, etc. Almost all the vocabulary (i.e., the values) of the primitive symbols N and f are single-morpheme words.[1] Also the particular f and the particular arguments which appear in the synonym-reducing CS (X *is synonym of* Y), and in the CS which are part of various ϕ_m (e.g., for producing *wh*-words out of

---

[1] There are cases of morphologically derived words which are syntactically and transformationally primitive, e.g., *a building*. But in most cases, a derived word in a sentence consists of a morpheme from a kernel sentence plus a trace of a transformation which had operated on that kernel sentence in forming the given sentence.

and), and in the metatype sentences, and in the metatoken sentences, and so on—all of these are special subclasses of words appearing in special members of N *is* N and other kernel-sentence forms.[2] The carrier sentences, which produce prime sentences out of the base operators, have other unique structures and vocabulary, different from those of the original kernel sentences.

With this correlation of structure and vocabulary goes a correlation of structure and meaning. Each subclass of words comprising a particular operator or argument class has a type of meaning fitting its syntactic relations (modality in ϕ_v; knowing, feeling, planning, etc., in ϕ_s; etc.). In N and f^1, i.e., in the kernel sentences, many, perhaps all, words (in the synonymless form) have unique extensions of co-occurrents (operators or arguments, respectively) with which they have normal acceptability.[3] In more fully describable and restricted subsets of language, such as the science sublanguages, the correlation of meaning with structure is sharper. This relation of meaning to structure is a special case of the relation of the meaning of any entity to the combinations in which it participates (see Chapter 6, fn. 3).

In the discussion of the f^2 types in 7.1.3 we saw examples of the kinds of meaning which can be borne by particular syntactic entities, i.e., by operators which appear in particular resultant positions in respect to particular operands. We have also seen more basic examples of how syntactic relations bear particular semantic effects: e.g., what makes certain sentence structures metalinguistic (5.4), or how individuation results from counting rather than from any specific linguistic element (5.7.2). We even obtain such general results as what is asserted in a sentence (its f^1, f^2, not its N or f^3). The ϕ-decomposition of particular types of sentence shows various facts: e.g., what assertions underlie the imperative $S!$ and question $S?$ (not $S.$ but *I ask whether S.*); or that conjunctions are binary predications (S_1 *because* S_2 asserts the causal relation, and can be transformed into a verb $S_2 n$ *causes* $S_1 n$); or that in the comparative, neither component sentence is asserted but only a relation of amounts (*He is taller than she is* $\leftarrow n$ *exceeds* m with conjoined sentences such as *His height (tallness) is n*, so

---

[2] The correlation is strengthened by the attempt to define the elementary operations in such a way that each morpheme appears in only one elementary operation; so that for example if *-ing* occurs in many transformations, it is because all these share a particular base operator as component.

[3] The corresponding hypothesis in structural linguistics is that difference in meaning between words correlates with difference between them in respect to their word neighborhoods. Transformational analysis permits a more precise formulation of the neighborhoods involved: they are the arguments and the operators.

that neither *He is tall* nor *She is tall* is asserted; of course, both of the latter may be false while the difference in amount may be true).

The *f*-operators under which a word occurs in a particular sentence also account for what seem to be changes in its meaning. For example, the difference in meaning between *to walk* and *a walk* is an interpretation of the fact that *a walk* occurs under a ϕ_v operator *take, have*. A more involved example is that of the peculiar English form *is to*, as in *The bell is to ring at three*. Syntactically, these sentences are odd because no auxiliaries can be added: ∄ *The bell will be to ring at three*. Semantically, they are odd because they carry the meaning of intention or of arranging for an outcome, even though the intender is not mentioned. It can be shown that these sentences are transformationally derived from a particular set of ϕ_s; these ϕ_s contain such verbs as *set, arrange*, and they insert *should* in their operand. Thus the sentence above is derived paraphrastically from *N set the bell that the bell should ring at three* → *N set the bell to ring at three*. The meaning of intention was brought into the sentence by the ϕ_s: *N set*. The *is* in *is to* is not a verb but the trace of a paraphrastic ϕ which zeroes the *N set* and inserts *be* to carry the tense, while *to* ← *should* obstructs further auxiliaries; and since paraphrastic ϕ do not change the meaning of their operands, the meaning of *N set*, which is reconstructible from the unique *is to*, remains even after it has been dropped.

Because of the connection between operator-structure and meaning, a sentence is ambiguous if its word-sequence can be produced by more than one ϕ, K sequences (5.1, 7.1.3). And the fact that resultants of particular ϕ are superclass sequences similar (at least initially) to those of other ϕ-resultants, and even of K (4.2.3.2), gives many ϕ-resultants a secondary grammatical meaning by the side of their direct meaning. The direct meaning is the cumulative one of their K and incremental ϕ, which includes both the meaning of the words of these components and the meaning of the positional relations these words have to each other within each component. The secondary grammatical meaning is the meaning that the word-position of the resultant has in the K (or other simplest structure) in which that positioning of word-classes first occurs. Thus in *The dog was seen by the boy*, the source K is *The boy saw the dog*, and the status of the boy as the actor ("subject" of *saw*) is preserved in the passive resultant. But the passive consists of a superclass sequence *N is A ...*, looking like a K such as *The dog was sick*, and therefore *the dog* has some secondary semantic property of being the topic of conservation (the "subject" of *was seen*) as it has unequivocally in *The dog was sick*.

We see in all this material that there is a connection not only between the ϕ, K decomposition and the meaning of a sentence, but also between kinds of syntactic connection and kinds of meaning. For example, the kinds of meaning which interpret the inequalities imposed by f^1 and $f^{1'}$ (called selection or co-occurrence restrictions in linguistics) are dictionary (lexical) meanings, while the kind of meanings due to the positioning of an operator's word-classes in respect to those of its operand are syntactic (grammatical) meanings. Almost everything that there is to say about the meaning of a sentence can therefore be obtained directly from the meanings and positions of the component ϕ, K. Hence, given this theory of base transformations there is little need for an additional semantic theory.

Within a subclass, i.e., in a fixed structural situation, there is an additional correlation of meaning with acceptance. It has been noted that all that is needed for establishing the transformations was a set of acceptance-inequalities. The data of the acceptance test are, however, stronger than this: for a given f_i^1, the inequalities are transitive, so that the N on which the f_i^1 operates are ordered. Since every syntactic fact has a semantic effect, we may ask what is the effect of this ordering, which is otherwise unused in the theory. The effect is that for each particular f_i^1 we can state relative semantic distance among its N arguments. It may turn out that these relative distances, summarized in S_{dict} (5.6.3) may determine the possibility of filling out the word repetition for a given S_1, S_2 conjoined by a C_s (7.4).

Many semantically special situations can be explicated by transformationally special situations. For example:

1. Marginal sentences are obtained by extending the domain of a transformation beyond its normal subclass of arguments. They can be formed for each transformation.

2. Metaphors: chiefly X_i in $S_1(N_1, X_i) \leftarrow S_1(N_1, X_{cli})$ as $[S_1(N_{ai}, X_{cli})$ consists in $S_1(N_{ai}, X_i)]$: *He planted his feet* $\leftarrow$ *He set his feet as one sets plants when planting them.* They do not accept some of the further ϕ which nonmetaphoric occurrences of the word accept. (X_{cli}: classifier of X_i; N_{ai}: appropriate to X_i.)

3. Jokes are produced in various transformationally stateable ways. For example, a joke is obtained when a marginal sentence is operated on transformationally as though it were of normal acceptability: as when to *He went out to meet his doom*, we add *But his doom did not meet him.*

4. Idioms are distinguished by the fact that the words in them have no local synonyms (*He kicked the bucket* $\neq$ *He kicked the pail*).

5. A is a paraphrase of B (in the synonymless S) only if A and B contain the same ordered kernel sentences and base incremental transformations. The paraphraseless set of sentences contains assertions only.

6. *A* is *n*-fold ambiguous only if there are *n* transformational paths leading to *A*, with at least two differences between any two paths. The later introduction of different transformational paths which do not lead to ambiguity, e.g., by allowing $\phi_i \phi_j = 1\phi_j$ when ϕ_j is not in the domain of ϕ_i, can be readily distinguished from the above. A kernel sentence is ambiguous only if it appears in two different acceptability graded subsets of sentences. We thus have a sharp distinction between grammatical ambiguity, above, and dictionary ambiguity.

7. Citing something (in a sentence) is found to consist of the inclusion in that sentence of an elementary sentence which pronouns and classifies the cited material.

8. Occurrence of something, as a token (in a sentence) of a given type, is found to be the pairing of the given type with a position in a sentence (or segment of a sentence) *A*, where *A* is cited in the sentence which asserts the occurrence.

9. Individuating reference is made only on nouns; in some structures only on counted nouns, which would be explained if, as seems to be the case, reference is made not to the nouns but to the operation of counting (which is carried out, at least linguistically, only on nouns).

10. What may be called performative operators are found to be zero-able, in a way that other operators of the same subset are not.

11. All sentences (except the nontransformable *Hello*, etc.) are derivable transformationally from assertions, so that the problem of understanding questions and commands and optatives is much simplified. Thus *Is he going?* is found to be a transformation not of *He is going.* but of *I ask you whether he is going or not.*

12. False sentences and false connections of sentences cannot be excluded by grammar, but meaningless connections between sentences can perhaps be excluded. Even the more subtle falsities in the use of language, as in advertising and propaganda, can be recognized by transformational analysis. For example, consider the sentence (from an advertisement): *To obtain a free gift mail us the label, and we will send your gift.* Here *your gift* has to be derived from *a gift which you have* or *a gift which has been assigned to you* (or: *associated with you*), etc. But the preceding *a gift* conflicts with this.

13. Paradoxes based on intensional impredicative definition of grammatical structures (as in *All sentences are false*) are perhaps not excludable by grammar, but paradoxes based on a certain form of impredicative sentences are isolated from the grammar of the rest of the language (i.e., are ungrammatical).

More broadly, the different parts of the abstract system have different systemic properties, not only semantic ones. Many individual rules appear as immediate consequences of the rules of the system, and many can be stated much more simply than in other formulations. For example, the rules of what may be zeroed are far simpler when stated in terms of the position of the antecedent in the operator than when stated in terms of neighboring words in the sentence. More important, the major defined entities have clear syntactic properties. The strings of 3.5, in contrast to constituents (3.4), have the property of being the least segments of sentences in terms of which there are no sentence-building operations at a distance from their operands. The transformational operators account for almost all morphology. The kernel sentences are finite, while the transformational resultants are denumerably infinite. Put somewhat differently: the f^1 are non-iterable, most f^2 are repeatable without bound, the f^3 are repeatable but in a way that soon falls off because the conditions for f^3 get used up (7.1.2). In the case of words which are restricted as to their co-occurrents, the number of normal co-occurrents for words of kernel sentences increases as long as our sample of the language increases, but restricted words (constants) of operators get no increase in co-occurrents past a certain size of sample: There is no definite limit to what words are normal as objects of *take* as kernel verb; but *take* as ϕ_v is normal with certain V including *walk* and not, for example, with *talk*, and no increase in sample will make it normal outside of a closed set.

The power added to language by specified structural entities, e.g., ϕ_v, ϕ_s, $N_{''}$, is great and is specific. These syntactic additions provide a power which is not only syntactic but also semantic. Somewhat similarly the syntactic possibility of adding $CS_{meta(i)}$ to S_i, i.e., of conjoining to a sentence other sentences that talk about it, brings into syntax some of the work that seemed before to lie outside it, because the additional meanings (grammatical, lexical, extralinguistic situational) which were needed in order to understand a sentence in a certain way are now available as syntactically distinguished parts of that sentence, enlarged. Somewhat similarly, also, the word-repetition condition for $SCS \ldots CS$ gives a syntactic explication of what a sentence is about, and the equivalence classes of discourse analysis show the structure for making a connected argument about something.

As a result of all these explicit properties, it is easy to construct additions and modifications of predictable character in language. Given the list of f^2 (7.1.2), it is a simple matter to construct additional subsets of f^2 having more arguments or dependencies among the arguments. The various constructions of Chapter 5 were examples of larger modifications of language.

There are various informational applications. The transformational decomposition of sentences (and also the synonym-removal of 6.5) gives each proposition a unique normal form, so that the elementary assertions and operators of the sentences of a text can be compared with each other and with those of related texts. The assertions and operations each carry fixed information (as seen above), and this makes possible an orderly survey and processing of the information in these texts. It turns out that in many cases, especially in argument as contrasted with narrative, the kernel sentences carry trivial information (relative to the discourse),[4] and that only the piling up of unary and binary operators makes the information important. The word-repetition condition for binary operators (and, less definitely, the equivalence-class condition for discourses) also makes possible a mechanical inspection of normalized texts to see if the conditions of meaningfulness are satisfied.

Since the trace of each operator is a specified addition or change at a specified point of its operand,[5] it is possible to inspect the sentence to see if the operator has acted in the sentence, and if so at what stage in the construction of the sentence. The decomposition (or decompositions) of each sentence into its partially ordered kernels and transformations is therefore computable and can be carried out by computer.[6]

The capacity for mechanically decomposing the sentences of whole scientific articles into a normal form of elementary assertions with operators on them makes it possible to carry out inspection and critique of the grammatical meaningfulness of the argument in the article and to process the information in the article in various ways; it also makes it possible to

---

[4] In some cases, the kernel sentences contain semantically empty words, as in *gland functions*, which can be shown to be linguistically dependent upon neighboring operators. In such cases, especially for particular sublanguages, enlarged independent macro-kernels can be defined from the sequences of dependent segments.

[5] And not simply at a specified point of the sentence; indeed, the further production of the sentence has not yet been determined at the time when the given operator acted.

[6] A transformational sentence-analyzer has been designed by A.K. Joshi; see his *String Representation for Transformations*, Transformations and Discourse Analysis Papers 58, University of Pennsylvania 1966; *Transformational Analysis by Computer*, Proceedings of NIH seminar on computational linguistics, Bethesda 1966; also Danuta Hiż and A. K. Joshi, *Transformational Decomposition*, Proceedings of IFIP international conference on computational linguistics, Grenoble 1967. A sentence analyzer using string analysis and based on a single scan of the sentence was devised by N. Sager and is in operation at New York University: N. Sager, Syntactic Analysis of Natural Language, *Advances in Computers* L. F. Alt and M. Rubinoff, Eds., 8 (1967) 153–188. A program based on the cycling automaton of 3.7 has been produced under the direction of I. D. J. Bross and is in use for retrieving information from medical reports; see P. A. Shapiro, Acorn, *Methods of Information in Medicine* 6 (1967) 153–162.

compare the information in various articles within the same field. Any processing of the information in scientific articles is impossible, beyond a very rough approximation, without transformational analysis. For example, this analysis indicates the order of ambiguousness of a sentence; and comparisons of the various normal forms of an ambiguous sentence with the normal forms of neighboring sentences (ambiguous or not) in the discourse usually suffice to determine which normal form of each ambiguous sentence is the relevant (intended) one for the given neighborhood. Furthermore, if a word in a sentence is a member of more than one synonym set (i.e., if it is homonymous), the determination of the synonym set to which this occurrence of the word belongs can be made on the basis of the other words in the same elementary sentences or in certain related neighboring elementary sentences (see 5.10, end). All this could be done by precisely defined operations.

Beyond this, one can analyze the argumentation in scientific discourse by transformational and discourse-analysis methods. And one can study the structure of science sublanguages—their word subclasses, the special sequences of these which make elementary sentences, the role in scientific discourse of the various types of unary operators, and the particular types of binary connectives.

An important result of the reduction to base transformations is that many languages, perhaps all, have the same structure in respect to the most essential properties: e.g., having elementary sentences formed by one or more sets of f^1 operating on a small set of word classes, having a few sets of base unary operations which preserve acceptability inequalities, and binary operations with a word repetition condition. The zeroing transformation ϕ_z operates with rather similar conditions in the various languages, permutations ϕ_p vary more, and the morphophonemic transformations ϕ_m are very different from language to language, as are also the phonemic distinctions and, even more so, the phonemic composition and semantic range of each morpheme. This graded similarity among languages suggests investigations differentiating the essential structure common to all languages from the secondary and even accidental features. This similarity also suggests possibilities of proceduralized translation, since it is a far simpler matter to translate the normal form of sentences of one language into the normal form in another (5.10), than to translate the actual sentences (in which the more diversified paraphrastic operations, ϕ_z, ϕ_p, ϕ_m, have acted).

Transformational analysis also has relations, as yet unstudied, with the change and development of language. The formula for analogic extension of transformations beyond their domain is similar to the formula for analogic language change. The use of inverse transformations is the back-formation known in linguistics. And when a detailed investigation is carried out into certain transformations, one comes upon a marginal part of the domain where it is clear that the transformation is currently in process of being established or of being seriously extended in the language.

These properties of transformations, however, do not necessarily reflect the history of the language. They may reflect incipient changes, which may or may not carry through; and they may reflect partial similarities in the semantic content of various sentence forms.

In any case, transformational analysis gives a set of descriptive—even if not developmental—stages for language, since f^1 can only be defined on N, f^2 only on f^1, f^3 only on f^2. And the very simplicity of this system, which surprisingly enough seems to suffice for language, makes it clear that no matter how interdependent language and thought may be, they cannot be identical. It is not reasonable to believe that thought has the structural simplicity and the recursive enumerability which we see in language. So that language structure appears rather as a particular system, satisfying the conditions of Chapter 2 and perhaps also bound by a history, which may evolve not only in time but also by specialization in science languages, and which is undoubtedly necessary for any thoughts other than simple or impressionistic ones, but which may in part be a rather rigid channel for thought.

Index*

*Symbols are indexed alphabetically following the regular entries. See p. 228.